The Atlas of the Island of Ireland

Mapping Social and Economic Patterns

Census Data

Mapping and Boundary Data

First published 2015
by All-Island Research Observatory (AIRO) and the
International Centre for Local and Regional
Development (ICLRD)

AIRO, Maynooth University, Maynooth, Co Kildare, Ireland
ICLRD, c/o Centre for Cross Border Studies, 39 Abbey
Street, Armagh, BT61 7EB, Northern Ireland

ISBN 13 9780992746612

All mapping by AIRO

Designed and produced by LSD Limited,
Greenwood Business Centre
430 Upper Newtownards Road
Belfast BT4 3GY

Printed by Nicholson & Bass, Co Antrim

All census data for Ireland has been made available
through the CSO Small Area Population Statistics
Source: Central Statistics Office,
website: www.cso.ie

All census data for Northern Ireland has been made
available through NISRA

Source: Northern Ireland Research and Statistics Agency,
website: www.nisra.gov.uk

National Mapping Agency - www.osi.ie

All Republic of Ireland boundary data have been made
available through the Ordnance Survey Ireland
© Ordnance Survey Ireland/Government of Ireland
Copyright Permit No. MP 005814

All Northern Ireland boundary data have been made
available through the Ordnance Survey of
Northern Ireland
© Crown Copyright 2014
Ordnance Survey of Northern Ireland
Permit No. 140029

Acknowledgements

The *Atlas of the Island* of Ireland has been jointly developed by the International Centre for Local and Regional Development (ICLRD) and the All-Island Research Observatory (AIRO). Prepared under the Evidence-Informed Planning element of the Cross-Border Spatial Planning and Training Network (CroSPlaN II) programme, this project is part-financed by the European Union's INTERREG IVA programme managed by the Special EU Programmes Body.

We are most grateful for the funding that was made available to support this project: the development of this hard back Atlas; the development of the on-line digital Atlas and the hosting of a series of census seminars and data days. We are also most grateful to the Department of the Environment, Community and Local Government (DELCG) for their financial assistance in co-funding the printing of this Atlas and funding the preparation of the All-Island Deprivation Index.

The *Atlas of the Island of Ireland* principally maps census data available in both Ireland and Northern Ireland. Two main organisations are responsible for gathering and disseminating this key demographic, social and economic information. The Central Statistics Office (CSO) in Ireland are responsible for the development of the Small Area Population Statistics (SAPS) and are available at both the Electoral Division (ED) and Small

Area (SA) geographical scale. We are particularly grateful to Deirdre Cullen, Gerry Walker, Aileen Healy and the staff at the Census Division for assisting us on this project and developing additional tables that were not available through SAPS. The Northern Ireland Statistics and Research Agency (NISRA) have provided all of the required statistical information for Northern Ireland. The transfer from raw census data to maps required the use of a variety of geographical boundary files. We are most grateful to the team at the Ordnance Survey Ireland and Land and Property Services Northern Ireland who assisted in administering the relevant copyright licences.

The development of both the mapping within this Atlas and the accompanying on-line digital atlas, an equally important aspect to this overall project, could not have been carried out without the assistance and dedication of the wider AIRO team. Sincere gratitude to Eoghan McCarthy who managed the development of the underlying data for this Atlas and designed the accompanying on-line mapping tool. Special thanks also goes to Aoife Dowling and Alan Campbell for their assistance in the production of many of the maps within this Atlas. This team have also been responsible for running an excellent series of census seminars and census 'data days' over the last 18 months. For more details on the accompanying on-line Atlas see www.airo.maynoothuniversity.ie. Special thanks also goes to

Clara Malone for diligently proof reading the many pages within this Atlas.

Our sincerest thanks to all those from across the island of Ireland who have contributed to the penning of this Atlas. Their analysis and insights into what Census 2011 is telling us about the 'state of play' across the island - and policy implications - is why this Atlas will be a useful tool for decision-makers, practitioners and academics alike.

The advice and assistance of the staff at the Centre for Cross Border Studies, as lead partner in this INTERREG-funded programme and in organising the series of census seminars and 'data days', has been invaluable. Special thanks also goes to the Border Regional Authority, Inter*Trade*Ireland, Land and Property Services NI and Dundalk Institute of Technology for hosting our 'data day' workshops.

Finally, the design and layout of the Atlas has been undertaken by Ashley Bingham at LSD Limited. Special thanks to Ashley for her patience, dedication and skill in bringing this project to a conclusion.

Mr. Justin Gleeson
Director, AIRO

Ms. Caroline Creamer
Acting Director, ICLRD

Foreword

An atlas is about curiosity and discovery and can help us understand how we are connected to each other. When we open an atlas, we learn more about where we live and what we have in common with our neighbours across town or across the border.

As the work of peace-building and socio-economic stabilisation continues, it is worth remembering the early days of cross-border cooperation on the island. An often lamented absence of cross-border data sets meant that those leading cross-border planning and cooperation became, by necessity, adept practitioners of the art of uncertainty, while simultaneously becoming the champions for a new genre of evidence and the growth of a skills culture across sectors which supported evidence-based working.

Since the International Centre for Local and Regional Development (ICLRD) was formed in 2006, compatible cross-border data mapping has been one of the core activities among the ICLRD partners. In 2008, the ICLRD and All-Island Research Observatory (AIRO) published *Atlas of the Island of Ireland - Mapping Social and Economic Change* to help visualise the social, economic and demographic trends and interactions across the island of Ireland. Our hope was that a better

understanding of the geography of everyday living and working could promote more informed cross-border cooperation.

With the publication of this latest all-island Atlas in both print and digital formats, the community of researchers, policy-makers and practitioners involved in shaping the growth and development of their communities and regions now have a enormous resource to enable and strengthen their cooperation. This includes cooperation between and within local government and other sectors whose involvement can enhance implementation of the *Framework for Co-operation: Spatial Strategies for Northern Ireland and the Republic of Ireland*, jointly published by the Department for Regional Development (Northern Ireland) and Department of Environment, Community and Local Government (Ireland) in 2013.

Evidence informed decision-making is now at the forefront of good planning and governance practices. Today the planning and public administration systems in Ireland and Northern Ireland are undergoing significant reforms. Now more than ever, we rely on good information, engaging communities and an understanding of their needs in shaping how services are delivered, where people live, how economies grow

and how communities promote long-term sustainable development and wellbeing. This 2015 all-island atlas can support this trend, and takes advantage of the tremendous leaps in digital and mapping technology by providing digital tools that are accessible online.

The *Atlas of the Island of Ireland* can be seen as a best practice in cross-border cooperation that can be adapted to other regions in Europe and beyond. We are extremely thankful for the consistent support of both Governments, the Special EU Programmes Body and the work of a network of specialists on both sides of the border led by Mr. Justin Gleeson of AIRO.

This atlas is a significant, accessible resource containing validated data which tells many stories about our communities and populations. These stories can change our understanding and give us choices as to how we can focus on planning our shared future resources in a way that is best for all. We invite you to use it and to share it with others.

Caitríona Mullan
ICLRD Chair

John Driscoll
Former Director of ICLRD

Authors

Justin Gleeson is the Director of the All-Island Research Observatory (AIRO) and is based at Maynooth University. He is a member of the International Centre for Local and Regional Development (ICLRD) and has led the development of a series of cross-border data projects in recent years with colleagues at ICLRD and the National Institute for Regional and Spatial Analysis.

Andy Pollak was the founding director of the Centre for Cross Border Studies (1999-2013). He was secretary of SCoTENS, the Standing Conference on Teacher Education North and South (2003-2013) and a board member of the International Centre for Local and Regional Development (2006-2013). He is a former Irish Times Belfast reporter, religious affairs correspondent, education correspondent and assistant news editor. He blogs at www.2irelands2gether.com

Professor James A Walsh has been a Vice-President of Maynooth University since 2005 prior to which he was Head of the Department of Geography. He was a founder member of the National Institute for Regional and Spatial Analysis (NIRSA) and of ICLRD. His research interests are in spatial planning, rural development and spatial aspects of demographic change in Ireland. He has co-authored a number of census based atlases including *People and Place: A Census Atlas of the Republic of Ireland* (2007).

Dr. Chris van Egeraat lectures economic geography at Maynooth University. He is a research associate of the National Institute for Regional and Spatial Analysis (NIRSA) and the Chair of the Regional Studies Association – Irish Branch. He has been involved in a series of cross-border research projects, including an exploration of the advantages of all-island sectoral ecosystems, for InterTradeIreland.

Gavin Daly works with ESPON, the European Observation Network for Territorial Development and Cohesion, based in Luxembourg. He works extensively in transnational and cross-border contexts and has led a number of major European studies on territorial development and spatial planning. He was formerly an advisor to the Irish government on planning policy, sustainable transport and climate change.

Professor Rob Kitchin is an ERC Advanced Investigator at the National Institute for Regional and Spatial Analysis (NIRSA) at Maynooth University and a Principal Investigator for The Programmable City project, the Digital Repository of Ireland (DRI), and the All-Island Research Observatory (AIRO). He is a member of the International Centre for Local and Regional Development and has worked on a number of cross-border data and spatial planning projects.

Professor Mark Boyle is the Director of the Maynooth University Social Sciences Institute at Maynooth University. He is Deputy Chair of the Irish Social Sciences Platform (ISSP) and a Board member of the International Centre for Local and Regional Development (ICLRD).

Dr. Andrew McClelland is an International Centre for Local and Regional Development (ICLRD) Research Associate who has worked on several ICLRD projects under the Cross-Border Spatial Planning and Training Network programmes (CroSPlaN I and II), including most recently in 2014 when employed as a Postdoctoral Researcher at Maynooth University.

Dr. Ronan Foley is a lecturer in Geography at Maynooth University and has worked on several ICLRD projects on accessibility to health care facilities on the island of Ireland. His research interests fall into two main camps. One interest is in the area of GIS and Health Care Planning, with particular attention on accessibility and utilisation, health care service planning, mental health and inequalities in general. The other core research interest is in the area of therapeutic landscapes, with a focus on water and place and the developing research area of healthy blue space.

Trutz Haase has been an independent Social & Economic Consultant since 1995. In this capacity, he has been responsible for the design and implementation of monitoring and evaluation frameworks for government programmes aimed at alleviating poverty, as well as developing resource allocation models to target social expenditure on the basis of objective need criteria. He is best known for his work on the development of the Haase-Pratschke Index of Relative Affluence and Deprivation. Work outside the Republic of Ireland includes studies for the Northern Ireland Statistics & Research Agency (NISRA), Special EU Programmes Body (SEUPB) and International Fund for Ireland (IFI).

Dr. Jonathan Pratschke is a Research Fellow at the Department of Economics and Statistics of the University of Salerno in Southern Italy, where he lectures on social inequalities and research methods. He is co-author of the Haase-Pratschke Index of Affluence and Deprivation and has published widely on the construction of composite indicators using aggregate-level data. His current research interests are centred on the analysis of social inequalities in health, education and well-being in European countries.

Contents

List of Figures

List of Tables

Conclusion

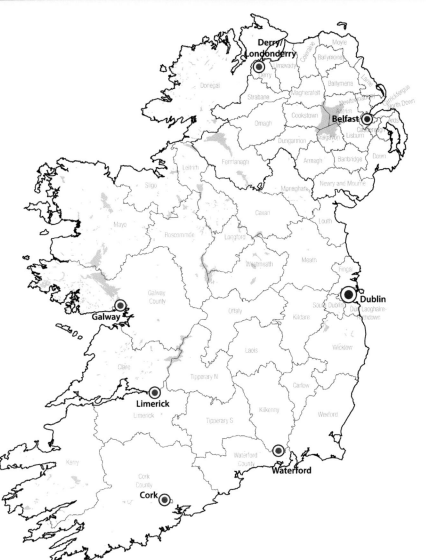

REFERENCE MAP
Local Authorities (Ireland) and
Local Government Districts (Northern Ireland)

Northern Ireland
Local Authorities
Water

Chapter 1

Introduction

Andy Pollak

The All-Island Research Observatory (AIRO) is one of a group of cross-border and all-island research organisations which emerged in the decade after the 1998 Belfast/Good Friday Agreement, the excellence of whose work has done much to put North-South cooperation on official agendas in both parts of Ireland. Among the others were the Centre for Cross Border Studies, the International Centre for Local and Regional Development (ICLRD), the Standing Conference on Teacher Education North and South (ScoTENS), the Institute for British Irish Studies and the Institute of Public Health in Ireland.

Between them these new bodies provided an invaluable source of research data and analysis for policy makers and planners, local authority leaders and others interested in working to minimise the practical barriers between Northern Ireland and Ireland for the mutual benefit of the people of the two jurisdictions. An impressive body of work has been produced in inter-governmental cooperation, spatial and environmental planning, local government, rural, urban and economic development, public service delivery (including shared services), education, health, housing, ethnic minorities and impact assessment. It is only disappointing that the administrations in Belfast and Dublin have not made better use of this wealth of research material to progress the cause of cross-border cooperation as part of the Northern Irish peace process.

The distinguished Northern business leader Sir George Quigley, who is widely acknowledged as the 'father' of contemporary North-South cooperation, has picked out the work of the ICLRD and AIRO for particular praise. In a 2013 interview, just before his death, he noted the former's work on mapping functional territories throughout the island and the latter's on mapping socio-economic indicators in the first, 2008 edition of *The Atlas of the Island of Ireland*. "These are potentially exciting concepts since, put at its simplest, they could hopefully be developed to provide guidance in an island context on what services should be put where, having regard to optimum catchment areas, thereby enhancing accessibility and ensuring that services are affordable, economically operated and effectively configured and managed to sustain high quality."

AIRO has been a pioneer in exploring ways of working on a practical, mutually beneficial way for the whole island of Ireland. The latest manifestation of this is the second *Atlas of the Island of Ireland*, with its unique presentation of social and economic information from more than 23,000 Small Area units throughout the island. Through this atlas, and the accompanying and highly accessible online mapping system[1], you will find harmonised and comparative data from the 2011 censuses in both jurisdictions on population, the labour force, economic activity, education, transport, housing, nationality and ethnicity, religion, health and caring, and affluence and deprivation.

One clear conclusion from reading this atlas is that there are far more similarities than differences between North and South. Both parts of the island have seen population growth unmatched anywhere else in Europe. The vast majority of workers in both are now employed in the service sector. A striking example of a similar trend is the overwhelming increase in both jurisdictions of car travel, with 82% of households in Ireland and 77% in Northern Ireland having at least one car or van. To the dismay of policy makers, the figures show a matching decline in use of other modes of transport, with, for example, a 6% decline in commuters walking or cycling to work between 2001 and 2011. The growth of dispersed commuter belts around all our cities has led to this, and

there must be serious questions about how government policies can do anything to reverse this overwhelming car dependency.

There are, of course, also significant differences. While a high proportion of people in both jurisdictions enjoy good health, Northerners' perception of their health is markedly worse. There has traditionally been much more social housing in Northern Ireland, although in the boom and crash of the period between the 1990s and 2007 both private housing sectors followed similarly unfortunate trajectories (the expansion in the South was far more dramatic and thus more damaging when it imploded: for example, the North has nothing to match the 'ghost estates' of the Upper Shannon region in Cavan, Leitrim, Longford, Roscommon and Sligo).

One area where there are striking and well-known differences is, of course, religion. In this section, however, there are also a few surprises. Across the island the *Other Religion* grouping has grown by over 200% since 1991, with the great majority in the South, reflecting higher immigration there. More surprising is the fact that although in absolute terms Dublin and Belfast have the highest absolute numbers in the No Religion category, the nine local authorities with the highest rates per head of population in this category are all in Northern Ireland, led by North Down and Carrickfergus. Richard Dawkins might see some reason for optimism here!

The highly regarded All-Island HP Deprivation Index created by Trutz Haase, Jonathan Pratchke and Justin Gleeson, is also updated in this new atlas, with, for the first time, a harmonised set of indicators measuring aspects of affluence and deprivation in an identical manner in Ireland and Northern Ireland (before there were significant difficulties in aligning data on education, social class and unemployment). Here the Central Statistics Office (CSO) has done extensive work to reclassify variables to match those in Northern Ireland. The authors' overall conclusions are that Northern Ireland is now more affluent than Ireland, with more extreme deprivation in the latter – a reversal of the position during the Celtic Tiger years as measured in the previous pilot version of the index.

This atlas is a milestone in many ways. Its genuinely all-island nature is a model for other people in the research and public policy arenas to follow. Its creator, AIRO, has worked alongside the two government statistics agencies, the CSO and the Northern Ireland Research and Statistics Agency (NISRA), to show how cross-border data – which in many vital areas such as trade and the economy remains seriously ill-matched - can be brought together to aid both mutual understanding and coordinated policy making. There are many areas of public policy which would benefit from more North-South

cooperation to bring improved services and cost savings on both sides of the border: health and hospital provision, research in third level institutions, public water and environmental services, to name but four. AIRO and this atlas point the way towards harmonised and attractively presented data sets to facilitate such necessary cooperation.

Alongside the atlas AIRO has produced simple and accessible on-line mapping tools to enable everybody from senior civil servants to people in the community to use data in the daily decisions that affect all citizens' lives – and has done this on an all-island basis. It has a particular commitment to working in the often forgotten region along the border, evidenced by the 'data days' it organises regularly in towns in that region to train local officials and community activists. The success of these 'data days' is underlined by the facilitation of these events for AIRO by other organisations such as the Border Region Authority, InterTradeIreland, Land

and Property Services (NI) and Dundalk Institute of Technology.

Finally, this is a pioneering European project at a time when the European Union often gets a bad press. It is funded by the EU INTERREG cross-border programme; has been developed by AIRO from an Maynooth University base with its cross-border partners, the International Centre for Local and Regional Development and the Centre for Cross Border Studies; and uses data from censuses which in Ireland and Northern Ireland, as in the rest of the EU, were held for the first time on the same day in 2011. It is no coincidence that AIRO is beginning to be recognised elsewhere in Europe as a highly innovative Irish initiative that has a lot to teach other and larger member states.

Andy Pollak was the founding director of the Centre for Cross Border Studies.

Chapter 2

Population Distribution and Change

Prof. James A Walsh

2.1 Introduction

The demographic profile of the island of Ireland in the early years of the twenty first century must be considered as an outcome from the different interactions of long term economic, social and political forces in both parts of the island, and also as an outcome from Ireland's engagement with the wider world, especially in terms of migration patterns. The total island population of just over 6.4 million persons in 2011 is the highest on record since the early 1850s when the total was 6.55 million, approximately 1.6 million less than the peak of almost 8.2 million ten years previously prior to the disastrous famine.

The demographic profiles of Ireland and Northern Ireland have evolved differently due to contrasts in the economic, social and political trajectories of both parts of the island. At the start of the twentieth century the population of the island of Ireland had fallen to under 4.5 million and declined further to 4.39 million persons in 1911 with 28.5% of the total residing in Northern Ireland. The proportions residing in both parts of the island one hundred years later in 2011 are broadly similar with 28.3% in Northern Ireland. However, over the intervening 100 years there was a net increase of just over two million persons (45.8%) in the island population with 72% of the increase in Ireland. This increase was accompanied by demographic trajectories that differed considerably.

The long-term decline in the population of Ireland continued at every census up to 1961 apart from a small increase of 5,000 in the immediate post World War period of 1946-51. The 1961 total of only 2.818 million was 322,000 (10.3%) less than in 1911. By contrast in Northern Ireland despite significant decline due to emigration between 1922 and the early 1930s there was a net increase of 176,000 (14.1%) up to 1961. In aggregate the total island population in 1961 was

at its lowest level of 4.246 million apart from in 1926 when the total was only 4.226 following very high net emigration of 165,000 during the turmoil of 1911-1926 from the part that later became the Republic of Ireland. By 1961 the Northern Ireland share of the total had increased to 33.6%. This contrast in trajectories reflected differences in social and economic conditions on both parts of the island with particularly high rates of net emigration from Ireland especially in the 1950s. Low rates of natural increase in the 1920s and 1930s coupled with comparatively high rates of net emigration were important factors in the earlier decades.

The late economic recovery in Ireland in the 1960s resulted in a net gain of 160,000 (5.7%) that cancelled out the entire net loss over the previous 35 years, but the more favourable economic context in Northern Ireland resulted in a further rapid expansion of 113,000 (8%) so that the share of the total island population residing in Northern Ireland reached 34.1%, the highest on record. By 1971 the total island population had recovered to just over 4.5 million.

The demographic trajectories of the 1970s and early 1980s were very different. More favourable economic conditions in Ireland were accompanied by drastically reduced net emigration, continuing high birth rates linked to increased marriage rates, and additionally natural increase was bolstered by declining death rates that were partially in response to the changed age profile following the net emigration of the decades prior to the 1960s. In Northern Ireland, by contrast, the political context had become seriously unstable which impacted adversely on the overall economic context. The population continued to grow but at a much slower pace than previously (increase of only 3,000 between 1971 and 1981 compared to 465,000 in Ireland over the same period). By 1981 the total had reached almost 5 million. With growth in Ireland continuing more rapidly the Northern Ireland share had fallen back to 30.8% of the total by 1986.

Over the quarter century between 1986 and 2011 the total island population increased by 1.289 million or 25% which was uniquely high in Europe. The more favourable economic context in Ireland was accompanied by high net in-migration and high birth rates. The 30% rate of increase in Ireland was double that in Northern Ireland and Ireland's share of the total exceeded the 1911 share for the first time (71.7% v 71.5%).

Within this overall pattern of change there have been significant adjustments in the structure (age profiles) and distribution of the population. Changes in the economy resulting in different employment patterns, social changes contributing to considerably increased female participation rates in the workforce, and widespread car ownership have resulted in increased urbanisation, expansion of commuter hinterlands especially around the largest urban centres, some counter-urbanisation and rural repopulation but also continuing population decline in extensive parts of the west and northwest. These changes are accompanied by increasing differentiation between areas with more youthful populations, less traditional household compositions, higher education attainment levels, and greater integration into the knowledge based economy, and on the other hand sparsely settled rural areas with ageing populations often living alone, where provision of public and private services and employment opportunities are contracting.

The remainder of this chapter focuses on the 2011 population and changes over the previous decade.

2.2 Population Density

Population densities throughout Northern Ireland have been consistently much higher than in Ireland. In 2011 the Northern Ireland density of 137 persons per square kilometre was double that of Ireland (67). The highest densities are, as expected, in the cities and the surrounding districts. Densities between 50 and 100 persons per sq. km occur in many of parts of Northern Ireland beyond the Belfast conurbation, but in Fermanagh and Moyle (north Antrim) the densities are only 36.1 and 34.5. In Ireland even lower densities ranging from 30.2 to only 21.2 occur in Galway county, Roscommon, Mayo and Leitrim. Such low densities which are much less than in most parts of western Europe, present significant challenges for the provision and maintenance of a wide range of both public and private services. Map 2.1 outlines that changes in densities occur very gradually except in localised areas of rapid population change.

2.3 Population Change 2001/02 to 2011

Over the decade to 2011 the total island population grew by almost 800,000 (14.25%), considerably faster than in any other European country. Natural increase was the dominant source of population increase in the early years of the decade. Following the accession of eight central and eastern European countries to the EU in May 2004 there was a surge in net migration (296,000) into Ireland between 2004 and 2008 making it by far the dominant force in population change. Northern Ireland also attracted large numbers of immigrants over this period (32,210), though on a lesser scale. The impact of

economic recession since 2008 has resulted in renewed levels of net emigration so that natural increase has again become the main influence on population change. In Ireland natural increase rose sharply from 24,800 in 2001 to 47,500 in 2011 due to a combination of a marginal increase in already high fertility rates and increased numbers of women in the child bearing age groups which is an outcome of the wave of immigration. The fertility rate in Northern Ireland also increased for similar reasons.

Over the decade the level of net migration into Ireland was almost twelve times the level into Northern Ireland where net migration contributed only 21% of the total increase in contrast to 47% in Ireland. This imbalance in the extent of population movements resulted in 84% (671,000) of the total increase occurring in Ireland.

The geography of population change between 2002 and 2011 in Ireland was very uneven with the most rapid increases in the peri-urban and longer distance commuter zones while decline continued in sparsely

settled and remote western coastal areas and also in some inland districts in the southwest and midlands (Map 2.2). Increases are also evident in towns located adjacent to the principal national roads and in coastal rural areas in the southeast and also in distinctive landscapes along the west coast.

At county level, increases between 37% and 40% occurred in Fingal, Meath and Laoighis followed by Cavan, Kildare, Longford and Wexford (Figure 2.1). Elsewhere there were significant increases in counties Cork and Galway associated with overspill from the city areas. While the influence of Dublin remains dominant, its impact became more widespread. Between 1991 and 2002, 47% of the total increase in Ireland occurred in Dublin and the three surrounding counties; between 2002 and 2011 the share had dropped to 33.6%. Conversely in an outer ring of eight counties there was a pattern of accelerated growth from 57,350 in the 1990s to 132,000 in the following decade. In contrast to the increases in peri-urban and other rural areas there were declines in the populations of Limerick and Cork cities

Figure 2.1: Population Change in Ireland, 2001/02 to 2011

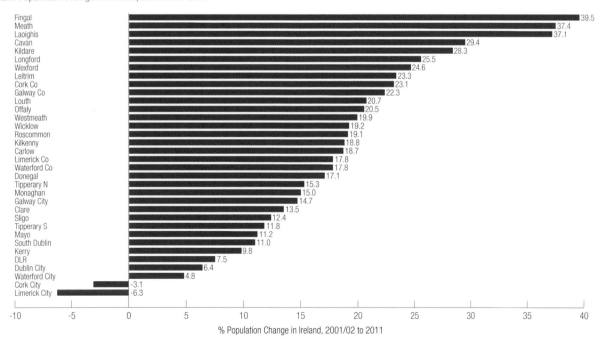

% Population Change in Ireland, 2001/02 to 2011

County	% Change
Fingal	39.5
Meath	37.4
Laoighis	37.1
Cavan	29.4
Kildare	28.3
Longford	25.5
Wexford	24.6
Leitrim	23.3
Cork Co	23.1
Galway Co	22.3
Louth	20.7
Offaly	20.5
Westmeath	19.9
Wicklow	19.2
Roscommon	19.1
Kilkenny	18.8
Carlow	18.7
Limerick Co	17.8
Waterford Co	17.8
Donegal	17.1
Tipperary N	15.3
Monaghan	15.0
Galway City	14.7
Clare	13.5
Sligo	12.4
Tipperary S	11.8
Mayo	11.2
South Dublin	11.0
Kerry	9.8
DLR	7.5
Dublin City	6.4
Waterford City	4.8
Cork City	-3.1
Limerick City	-6.3

Figure 2.2: Population Change in Northern Ireland, 2001/02 to 2011

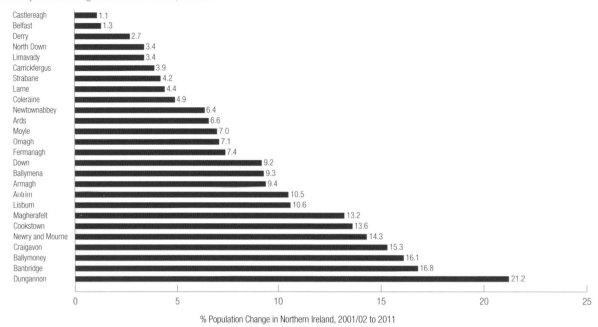

% Population Change in Northern Ireland, 2001/02 to 2011

and only marginal increases in the Waterford and Dublin city areas.

In Northern Ireland there are also pronounced urban-rural differences in the pattern of change contributing to a major east/west contrast. The highest growth rate was 21.2% in Dungannon district mostly due to net in-migration, followed by rates in excess of 13% in Banbridge, Ballymoney, Craigavon, Newry and Mourne, Cookstown and Magherafelt (Figure 2.2). In Ballymoney, Banbridge and Craigavon approximately half of the growth was due to natural increase with in-migration from other parts of Northern Ireland also important in Banbridge and Ballymoney while in Craigavon most of the immigrants are from outside Northern Ireland. Differences in fertility rates over a long period account for some of the difference in population increase between Newry and Mourne (14.3%) and North Down (3.4%). There were much smaller increases in eastern urban districts with older age profiles and also in Derry which has been impacted by high net out-migration. In contrast to Ireland there are very few extensive areas of rural population decline.

2.4 Age Profiles
There are significant differences in age profiles between the populations of both jurisdictions on the island of Ireland, and also within each jurisdiction. Ireland has a higher proportion of young persons aged under 14 years, 21.4% vs 19.6% in Northern Ireland, and conversely fewer aged 65 years and over, 11.7% vs 14.5%. The greatest differences are in the 30-39 age cohort with considerably higher proportions in Ireland that can be associated with the high birth rates of the 1970s that were augmented by net immigration in later years. The demographic 'echo' from this cohort is evident in the higher proportions of children in the population of Ireland in 2011 (Figure 2.3). A further contrast over the decade to 2011 was the reduction in the 15-24 years cohort in Ireland which can be attributed to the decline in births in the late 1980s and early 1990s, and additionally to very high emigration of young entrants to the workforce, especially in the later years of the decade as a consequence of the decline in the economy. In Northern Ireland the significant decline in the size of the 30-39 years cohort can be linked to fewer births as a result of higher emigration in the period of serious political disturbances in the 1980s; the echo from this

period is also evident in the decline between 2001 and 2011 in the share of the population aged 10-19 years (Figure 2.3).

Maps 2.3 to 2.7 illustrate the differences in age profiles across the island of Ireland. Major contrasts are evident between rural and urban areas especially in the 15-24 age cohort as young people leave rural areas for further education and employment. Rapidly growing urban centres with large commuter populations tend to have above average proportions in the working age groups which are also accompanied by above average proportions of children aged under 15 years. By contrast rural areas and towns beyond the commuter zones tend to have higher proportions of elderly populations reflecting successive waves of emigration (Table 2.4).

The average age in Ireland in 2011 was 36.1 years, an increase of two years from 1996. The adjustment can be attributed to increases in the numbers of middle aged persons and simultaneously the decline in the early working age groups as a consequence of increased emigration. The youngest populations are in Fingal, Kildare, Meath, South Dublin and Laois where the

Figure 2.3: Population pyramid of the island of Ireland, 2001/02 and 2011

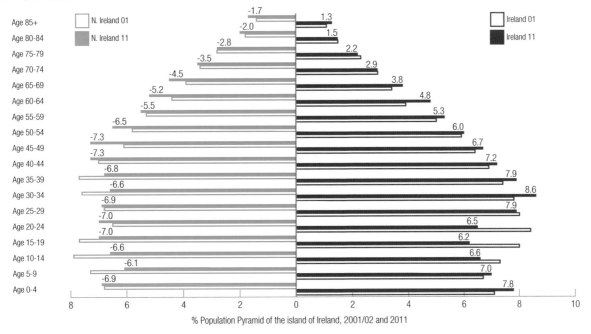

% Population Pyramid of the island of Ireland, 2001/02 and 2011

Table 2.4: Percentage Distribution of Persons in NUTS III regions, 2011

Region	0-14	15-24	25-44	45-64	65 plus	Total
Belfast (NI)	17.4	16.6	29.0	22.5	14.6	100
Outer Belfast (NI)	18.9	12.9	26.6	25.6	16.0	100
East (NI)	19.8	12.8	27.1	25.2	15.1	100
North (NI)	19.9	14.4	27.1	24.6	13.9	100
West and South (NI)	21.3	13.9	28.0	23.6	13.2	100
Border (Ire)	22.6	12.3	29.2	23.3	12.6	100
Dublin (Ire)	19.3	13.6	34.9	21.3	10.9	100
Mid-East (Ire)	24.3	12.0	32.6	22.1	9.0	100
Midland (Ire)	23.5	12.2	30.6	22.5	11.2	100
Mid-West (Ire)	21.3	12.9	29.6	23.8	12.5	100
South-East (Ire)	22.1	12.0	29.6	23.8	12.6	100
South-West (Ire)	20.9	12.3	30.6	23.6	12.5	100
West (Ire)	20.9	12.4	29.8	23.8	13.0	100
Northern Ireland	19.6	13.9	27.5	24.4	14.6	100
Ireland	21.3	12.6	31.6	22.7	11.7	100
All-Island	20.9	13.0	30.4	23.2	12.5	100

average ages are between 32.9 and 34.3. The inclusion of Laois in this group is partially a consequence of the extension of the Dublin commuter zone into the county since the late 1990s. By contrast the highest average ages are in the very rural counties of Leitrim, Mayo, Roscommon, Sligo, Kerry and Tipperary. The South Dublin population also has a high average age at 34.1 years reflecting the extensive tracts of inner suburbs in this part of the metropolitan area.

In Northern Ireland the population is older in many eastern districts which may be linked to lower fertility rates over many decades among the Protestant populations in contrast to higher rates, until recently, among the Catholic population. The youngest populations are in Newry and Mourne, Dungannon, Magherafelt and Cookstown. In Dungannon at least,

significant immigration over the last decade may be an important factor.

2.5 Dependency Rates

Maps 2.8 to 2.10 illustrate the variations in demographic dependency ratios. The young dependency ratio, measured as the ratio of children aged 0-14 years to the population aged 15-64 years, declined very significantly in both jurisdictions between 1971 and 2011 after which there was very little change. In 2011 the ratio is higher in Ireland at 31.9% compared to 29.9% in Northern Ireland. The highest ratios are in the most rapidly growing areas, especially in the hinterlands of the largest cities while the lowest ratios are found within the legally defined administrative boundaries of the five largest cities where the ratios vary from 20.9% in Cork city to 26.0% in both Limerick and Belfast.

The older dependency ratios defined as persons aged over 65 as a % of those aged 15-64 range from 17.4% in Ireland to 22.3% in Northern Ireland. The highest ratios are in the eastern districts of Northern Ireland where the ratios range from 26.5% in Larne to 28.7% in north Down. By contrast the lowest rates are in the Dublin commuter belt and also in Galway city. When the young and old dependency ratios are combined the lowest is in Galway city (34.9%) followed by Dublin, Cork and Limerick cities and Fingal county. By contrast the highest combined dependency ratios are mostly in the west and northwest (Mayo, Roscommon, Leitrim, Donegal) while in Northern Ireland the highest ratios are mostly in the east reaching a peak of 56.4% in Moyle in northeast Antrim.

MAP 2.1

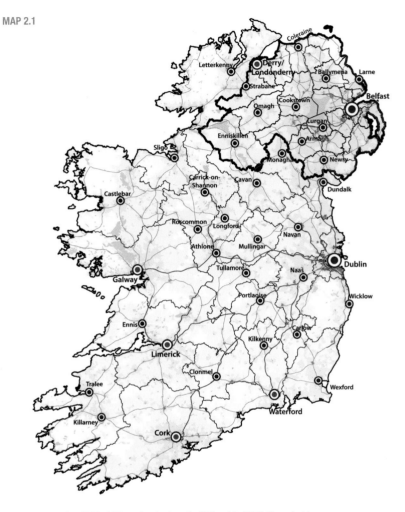

This map is part of an All-Island Atlas project developed by AIRO and the ICLRD. The project is part-financed by the European Union's INTERREG IVA programme managed by the Special EU Programmes Body.

Population Density per sq km, 2011
Small Areas (SAs)

Dublin City

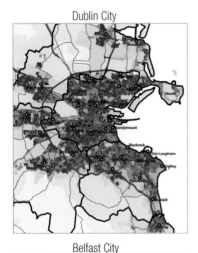

Belfast City

Population Density
per sq km (Small Areas)

0 to 10	1,001 to 2,000
11 to 30	2,001 to 4,000
31 to 100	4,001 to 10,000
101 to 300	10,001 to 35,000
301 to 500	> 35,000
501 to 1,000	

Northern Ireland
Local Authorities
Water

Motorways
Trunk/Primary Roads
Secondary Roads
Streets

MAP 2.2

This map is part of an All-Island Atlas project developed by AIRO and the ICLRD. The project is part-financed by the European Union's INTERREG IVA programme managed by the Special EU Programmes Body.

Percentage Population Change, 2001/02 - 2011
Small Areas (SAs)

Dublin City

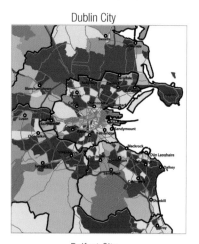

Belfast City

Data Source: Central Statistics Office (CSO), Northern Ireland Statistics and Research Agency (NIRSA)

% Population Change 01/02 to 11

- Population Decrease
- 0% to <10%
- 10% to <30%
- 30% to <60%
- 60% to <100%
- Greater than 100%

- Northern Ireland
- Local Authorities
- Water

- Motorways
- Trunk/Primary Roads
- Secondary Roads
- Streets

MAP 2.3

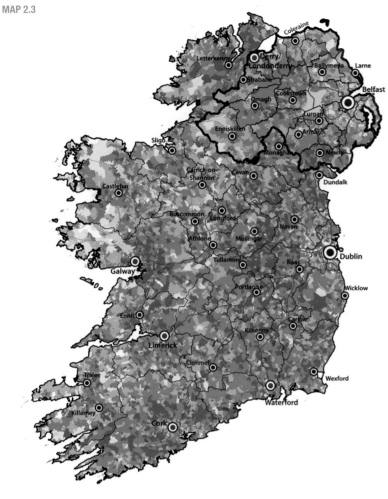

Percentage of population aged 0-14 years, 2011
Small Areas (SAs)

Dublin City

Belfast City

This map is part of an All-Island Atlas project developed by AIRO and the ICLRD. The project is
part-financed by the European Union's INTERREG IVA programme managed by the Special EU Programmes Body.

% Population
Aged 0 to 14 Years

Less than 10%
10% to <15%
15% to <20%
20% to <25%
25% to <35%
Greater than 35%

Northern Ireland
Local Authorities
Water

Motorways
Trunk/Primary Roads
Secondary Roads
Streets

Ordnance Survey
Ireland/Government of Ireland
Copyright Permit No. MP 005814

Crown Copyright 2014
Ordnance Survey of Northern Ireland
Permit No. 140029

Data Source: Central Statistics Office
(CSO), Northern Ireland Statistics and
Research Agency (NIRSA)

MAP 2.4

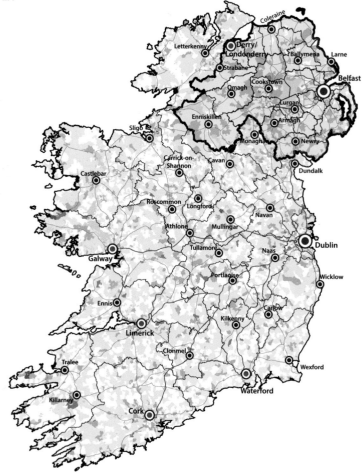

Percentage of population aged 15-24 years, 2011
Small Areas (SAs)

Dublin City

Belfast City

This map is part of an All-Island Atlas project developed by AIRO and the ICLRD. The project is
part-financed by the European Union's INTERREG IVA programme managed by the Special EU Programmes Body.

% Population
Aged 15 to 24 Years

Less than 10%
10% to <15%
15% to <20%
20% to <30%
30% to <60%
Greater than 60%

Northern Ireland
Local Authorities
Water

Motorways
Trunk/Primary Roads
Secondary Roads
Streets

Ordnance Survey
Ireland/Government of Ireland
Copyright Permit No. MP 005814

Crown Copyright 2014
Ordnance Survey of Northern Ireland
Permit No. 140029

Data Source: Central Statistics Office
(CSO), Northern Ireland Statistics and
Research Agency (NIRSA)

MAP 2.5

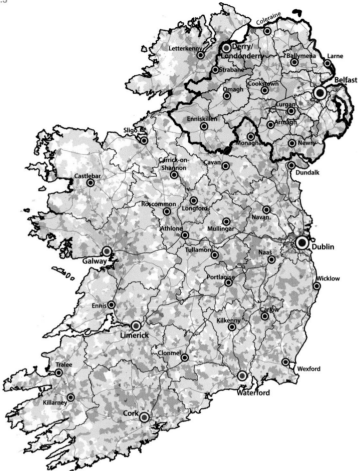

Percentage of population aged 25-44 years, 2011
Small Areas (SAs)

Dublin City

Belfast City

This map is part of an All-Island Atlas project developed by AIRO and the ICLRD. The project is
part-financed by the European Union's INTERREG IVA programme managed by the Special EU Programmes Body.

% Population
Aged 25 to 44 Years

Less than 22%
22% to <29%
29% to <37%
37% to <47%
47% to <60%
Greater than 60%

Northern Ireland
Local Authorities
Water

Motorways
Trunk/Primary Roads
Secondary Roads
Streets

Ordnance Survey
Ireland/Government of Ireland
Copyright Permit No. MP 005814

Crown Copyright 2014
Ordnance Survey of Northern Ireland
Permit No. 140029

Data Source: Central Statistics Office
(CSO), Northern Ireland Statistics and
Research Agency (NIRSA)

MAP 2.6

Percentage of population aged 45-64 years, 2011
Small Areas (SAs)

Dublin City

Belfast City

This map is part of an All-Island Atlas project developed by AIRO and the ICLRD. The project is part-financed by the European Union's INTERREG IVA programme managed by the Special EU Programmes Body.

% Population
Aged 45 to 64 Years

Less than 11%
11% to <18%
18% to <23%
23% to <28%
28% to <33%
Greater than 33%

Northern Ireland
Local Authorities
Water

Motorways
Trunk/Primary Roads
Secondary Roads
Streets

MAP 2.7

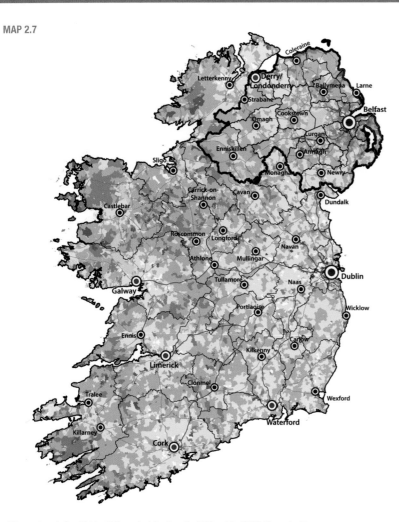

Percentage of population aged 65 years plus, 2011
Small Areas (SAs)

Dublin City

Belfast City

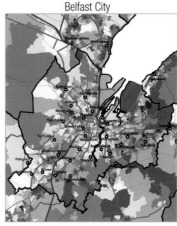

This map is part of an All-Island Atlas project developed by AIRO and the ICLRD. The project is
part-financed by the European Union's INTERREG IVA programme managed by the Special EU Programmes Body.

% Population
Aged 65 plus Years

Less than 6%

6% to <12%

12% to <18%

18% to <24%

24% to <35%

Greater than 35%

Northern Ireland

Local Authorities

Water

Motorways

Trunk/Primary Roads

Secondary Roads

Streets

MAP 2.8

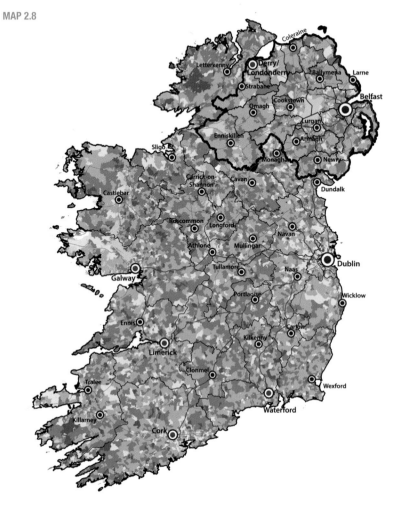

Dependency Rate, 2011 ((0-14 + 65plus)/(15-64))
Small Areas (SAs)

Dublin City

Belfast City

This map is part of an All-Island Atlas project developed by AIRO and the ICLRD. The project is
part-financed by the European Union's INTERREG IVA programme managed by the Special EU Programmes Body.

Dependency Rate
(Pop <15 + 65p) as % of Pop 15-64

Less than 28%
28% to <45%
45% to <57%
57% to <73%
72% to <110%
Greater than 110%

Northern Ireland
Local Authorities
Water

Motorways
Trunk/Primary Roads
Secondary Roads
Streets

Ordnance Survey
Ireland/Government of Ireland
Copyright Permit No. MP 005814

Crown Copyright 2014
Ordnance Survey of Northern Ireland
Permit No. 140029

Data Source: Central Statistics Office
(CSO), Northern Ireland Statistics and
Research Agency (NIRSA)

MAP 2.9

Youth Dependency Rate, 2011 ((0-14)/(15-64)) Small Areas (SAs)

Dublin City

Belfast City

This map is part of an All-Island Atlas project developed by AIRO and the ICLRD. The project is part-financed by the European Union's INTERREG IVA programme managed by the Special EU Programmes Body.

Youth Dependency Rate
(Pop <15) as % of Pop 15-64

- Less than 15%
- 15% to <25%
- 25% to <33%
- 33% to <42%
- 42% to <56%
- Greater than 56%

- Northern Ireland
- Local Authorities
- Water

- Motorways
- Trunk/Primary Roads
- Secondary Roads
- Streets

Ordnance Survey
Ireland/Government of Ireland
Copyright Permit No. MP 005814

Crown Copyright 2014
Ordnance Survey of Northern Ireland
Permit No. 140029

Data Source: Central Statistics Office
(CSO), Northern Ireland Statistics and
Research Agency (NIRSA)

MAP 2.10

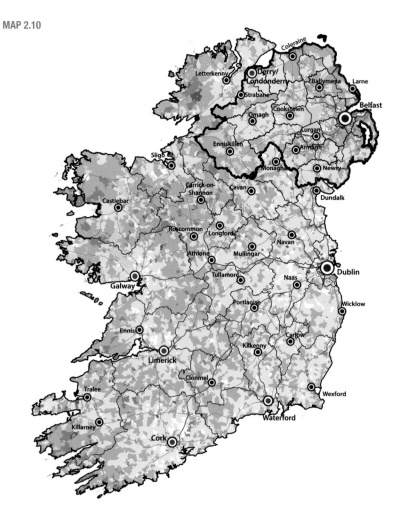

Old Dependency Rate, 2011 ((65+)/(15-64))
Small Areas (SAs)

Dublin City

Belfast City

This map is part of an All-Island Atlas project developed by AIRO and the ICLRD. The project is part-financed by the European Union's INTERREG IVA programme managed by the Special EU Programmes Body.

Old Dependency Rate
(Pop 65+) as % of Pop 15-64

	Less than 11%
	11% to <21%
	21% to <34%
	34% to <55%
	55% to <102%
	Greater than 102%

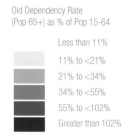

	Northern Ireland
	Local Authorities
	Water

	Motorways
	Trunk/Primary Roads
	Secondary Roads
	Streets

Chapter 3

Economic Status and Labour Force

Dr Chris Van Egeraat and Justin Gleeson

3.1 Introduction

This chapter will examine a number of aspects of the labour force within the island of Ireland, focussing on the spatial aspects of labour force participation and unemployment. The labour force is the part of the population aged 15/16-74 that is available for work. It includes both the employed and the unemployed but excludes students, retirees, homemakers and the unemployed that are not actively looking for work. Labour force dynamics are obviously influenced by natural population growth, age structure (see Chapter 2) and migration (see Chapter 8). Another important factor is the labour force participation (LFP) rate. This is the ratio between the labour force and the total population aged 15/16-74. If the proportion of the population that is not available for work increases, the LFP rate decreases and the labour force shrinks. The unemployment rate is the share of the labour force that is unemployed but actively looking for work. Since the unemployment rate is expressed as a percentage of the labour force we can see that the LFP rate also influences the unemployment rate.

Before presenting a detailed exploration of the labour force and its geography on the island of Ireland, we start with a note on methodology. In Census 2011, the data captured on economic activity, that is participation rates, employment and unemployment, were based on different methodologies in both jurisdictions. The census is Ireland is based on Principal Economic Status (PES)[2] whereas the census in Northern Ireland is based on the International Labour Organisation (ILO)[3] measure. Both measures produce very different results and, as such, a direct comparison of both is not possible. In order to facilitate a comparison between both jurisdictions, the

Central Statistics Office (CSO) undertook a conversion exercise to convert the data for Ireland to a comparative ILO basis[4]. Resulting from this, the following variables are now available on an all-island basis at the Small Area (SA) scale and will be discussed in the following sections:

- ILO Labour Force Participation Rate (LFP) - number of persons in the labour force as a percentage of the working age population (15/16-74)
- ILO Unemployment Rate - persons who, in the week before the survey, (a) were without work, (b) were available for work within the next two weeks and (c) had taken specific steps in the preceding four weeks to find work. The data include all persons in the 16 - 74 year age group who met these criteria.

3.2 Labour Force and Labour Force Participation Rate

Historically, the two parts of the Island have experienced very different labour force dynamics. During the 1980s, the labour force of Ireland grew only marginally, increasing by about 110,000 between 1981 and 1991. Natural population growth was offset by high levels of emigration and the LFP rate was low. This all changed significantly with the onset of the Celtic Tiger in the 1990s and first half of the 2000s. Strong natural growth of the population at working age, net migration following the accession of eight central and eastern European countries to the EU, and rising LFP rates translated into an unprecedented growth of the labour force (Figure 3.1). Between 1991 and 2007 the labour force nearly doubled from 1.15 to 2.28 million.

The LFP rate in Ireland increased significantly from a low of 53% in 1981 to a high of 64.7% in 2007. Much of the change in the 1980s and 1990s was a result of a

major shift in the female employment participation rates which increased from 30% to 47%. Increasing cost of living and housing contributed to accelerating the trend towards dual-income households.

The situation turned around sharply again with the onset of the global economic and financial crisis. In spite of a natural increase of the population at working age, the labour force contracted again by 144,000 (6.3%) between 2008 and 2014. Much of this can be explained by renewed high net emigration (estimated at approximately 80,000 between 2009 and 2014). Another contributing component was the change in the LFP rate which dropped to 60% in 2014 (Figure 3.2). A reduction in the employment prospects tends to be associated with a reduction in LFP rates. Some workers decide to work in the home and a larger share of the potential labour force may go into full-time education. In Ireland the latter was particularly the case for the age cohort 15-24 which witnessed a drop in the LFP rate from 56% in 2007 to 37% in 2014.

The labour force dynamics of Northern Ireland have been less intense. The labour force in Northern Ireland gradually increased during the 1980's from 663,000 in 1984 to 691,000 in 1991, an increase of 4%. After a slight contraction during the early 1990s, the labour force again expanded gradually during the remainder of the decade. Between 1994 and 2001 the labour force expanded by 62,000 to 753,000, an increase of 9%. This was mainly driven by natural growth of the population at working age. Compared to Ireland, the labour force trends between 1992 and 2008 were less influenced by LFP rate dynamics. The LFP rate fluctuated somewhat in the region of 58% to 60%. This changed with the onset of the global economic crisis in 2008. In sharp contrast to the situation in Ireland, the LFP rate

---

2   *PES: Respondents ticked one of a range of options to indicate whether they were at work, unemployed, a student, homemaker, retired etc.*
3   *ILO: respondents must satisfy certain conditions before they are deemed to be unemployed*
4   *For details on this conversion exercise see Appendix 3, Census 2011 Ireland and Northern Ireland at www.cso.ie/census*

Figure 3.1: Labour Force in Ireland and Northern Ireland, 2002 to 2013

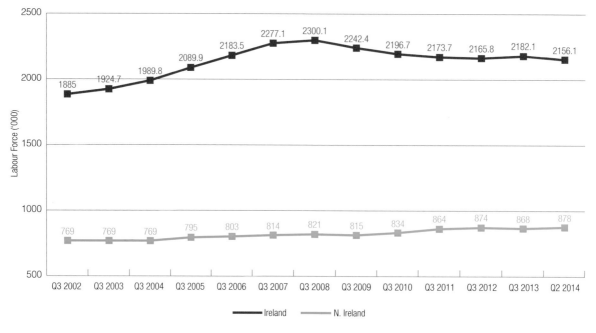

Figure 3.2 Labour Force Participation Rate, 2002 to 2013

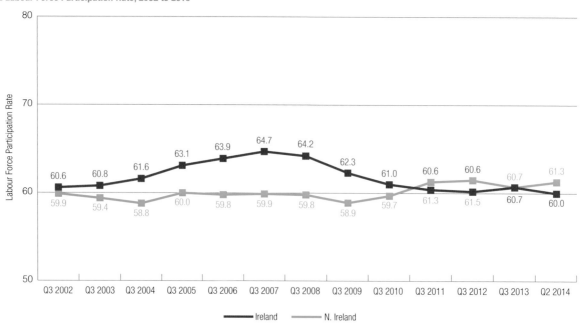

in Northern Ireland has increased to over 61% in 2014. This is reflected in the labour force that more or less continued its gradual growth, even after the onset of the economic crisis (Figure 3.2).

In 2011 the LFP rate in Northern Ireland (61.3%) was marginally higher than the rate in Ireland (60.4%) (see Map 3.1). Not only is there a clear Ireland/Northern Ireland differentiation but there are also distinct regional variations with areas of particularly low participation rates in southwest Cork, coastal parts of Kerry and Mayo, Leitrim and large parts of Donegal. The pattern is less clear in Northern Ireland but in general rates are lower in peripheral parts of the jurisdiction - both along the border and in the more rural inland areas. In Ireland the lowest rates (<58%) are found in Cork City, Donegal, Leitrim and Kerry with the highest rates (>66%) observed in Fingal, Meath, Kildare and South Dublin. In Northern Ireland the lowest rates (<57%) are in Moyle, Newtownabbey, Cookstown, Strabane and Belfast with highest rates (>68%) in Antrim, Ballymoney, Banbridge and Carrickfergus.

There are also distinct patterns in both the Dublin and Belfast hinterlands. In Dublin, the highest rates are found in the south city core and in many of the peripheral locations of the city such as Sandyford, Tallaght, Lucan, Blanchardstown and Swords. Many of these areas have undergone a significant transformation in recent years with high levels of new housing catering for economically active populations. Inside the M50 corridor, lower rates are evident in large parts of Dún Laoghaire-Rathdown and north Dublin City, generally reflecting elderly populations and higher concentrations of full time students. In Belfast there is also a very clear spatial pattern with highest rates in eastern Belfast (>70%) and much lower rates in the central parts of the city (probably picking up large student populations near Sranmills) and large parts of west Belfast (Twinbrook, Lower Falls, Shankill etc.).

3.3 ILO Unemployment Rate

To be classed as unemployed you must satisfy three conditions: you were not working the previous week, you are available for work, and you are looking for work. As of Q2 2014 11.8% (254,500) of the labour force in Ireland and 6.7% (59,000) of the labour force in Northern Ireland were unemployed.

In spite of different trends in the underlying components, the historic unemployment trends in Ireland and

Northern Ireland are broadly similar, until the most recent economic crisis. In Ireland, the opening up of the economy since the end of the 1950s and subsequent accession to the EEC resulted in strong employment growth and a concomitant fall in unemployment rates during the 1960s and early 1970s. However, the open character of its economy also meant that Ireland was disproportionately affected by the global economic crisis of the 1980s, leading to falling employment and strongly rising unemployment, reaching a high of 17% in the mid-1980s. This, in spite of the fact that the labour force itself was growing only marginally, due to rising emigration and the persistently low labour force participation rate described in the previous section. Unemployment rates dropped somewhat in the second half of the 1980s but rose again in the early 1990s. In Northern Ireland unemployment rates were rising in the early 1970s due to the intensification of the Troubles. The economic crisis of the 1980s magnified the problem, resulting in unemployment rates on par with Ireland. Similar to the situation in Ireland, unemployment dropped in the second half of the 1980s and rose again in the early 1990s.

After this, the fortunes changed in both jurisdictions. In Ireland, the economic activity during the Celtic Tiger era had no problem absorbing the spectacularly growing labour force. Rising employment in the foreign companies, services and construction led to virtually full employment in the early 2000s (Figure 3.3). The ILO unemployment rate fluctuated from 4.4% in 2002 to 4.7% in 2007. Likewise, in Northern Ireland unemployment rates gradually came down from 12.1% in 1992 to 4.7% in 2007.

The onset of the global economic and financial crisis in 2007 led to rising unemployment in Ireland and Northern Ireland. However, the unemployment trends in the two jurisdictions differed substantially in this period. In Ireland, in spite of a contracting labour market, the unemployment rate increased three-fold to a high point of 15.1% (328,100) by mid 2011. The declining LFP rates during this period would suggest that the "real unemployment" rate was substantially higher than that. The unemployment rate in Northern Ireland has been a lot more stable, peaking at a 'mere' 7.8% in 2012 (68,000), in spite of the expanding labour force discussed in the previous section.

One of the major factors in the large increase in the rate of unemployment in Ireland was the collapse in the

construction industry. In 2006, the construction industry in Ireland employed 155,774 people, representing 9.6% of those at work. By 2011 this figure had decreased by -44%, to 87,371, representing only 4.8% of those at work. The construction industry in Northern Ireland on the other hand did not suffer such a collapse and has remained at a buoyant rate of 8.2% in 2011.

After peaking in 2012, unemployment in Ireland is falling sharply again to 11.7% in Q2 2014. However, this remained the eighth highest unemployment rate of all EU28 countries (Figure 3.4). Only Greece (26.9%), Spain, Croatia, Cyprus, Portugal, Slovakia and Italy had higher rates. In comparison, Northern Ireland had the eighth lowest rate (6.7%), just ahead of the UK as a whole.

An additional worrying aspect to unemployment in Ireland is long-term unemployment rate, defined as the percentage of unemployed people who have been out of work for more than a year. By late 2012, 62.5% of those unemployed in Ireland were long-term unemployed – the fourth highest rate among the EU28 countries. Again, this is far higher than the rate of 48% observed in Northern Ireland. Interestingly, with the exception of recent post-crash years, Northern Ireland has continually had a substantially higher level of habitual unemployment (Figure 3.5).

Census 2011 enables a detailed illustration of the scale and distribution of unemployment across the island at the depths of the economic recession (see Map 3.2). The Ireland/Northern Ireland divide is very clear, with a drop off in rates immediately north of the border. Within Northern Ireland unemployment is higher along border areas with the highest rates observed in the districts of Derry (12.4%), Strabane (11.8%), Limavady (11.3%) and Newry and Mourne (9.3%). Lowest rates in Northern Ireland were recorded in the east of the country and predominantly in areas within close proximity to Belfast such as North Down (5.5%), Antrim (5.5%), Newtownabbey (5.5%) and Castlereagh (4.7%).

Unemployment rates are consistently higher in Ireland. All local authorities, with the exception of Dún Laoghaire-Rathdown (11.3%) and Sligo (11.7%), have rates that are higher than any local government district in Northern Ireland. Highest rates were recorded in Limerick City (27.8%), Clare (19.6%), North Tipperary (19.5%), Waterford City (19%) and Wexford (18.9%). Some of these local authorities are suffering from a relatively traditional industrial structure, characterised

Figure 3.3: Unemployment Rate in Ireland and Northern Ireland, 2002 to 2014

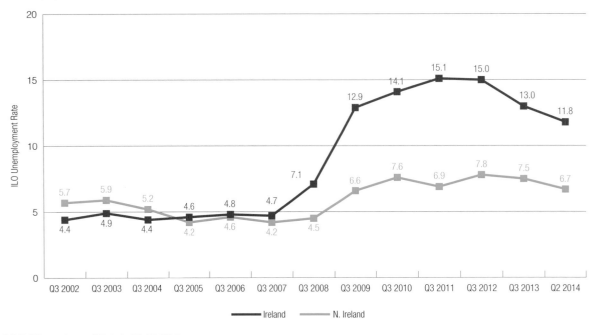

Figure 3.4: ILO Unemployment Rate in EU, Q2 2014

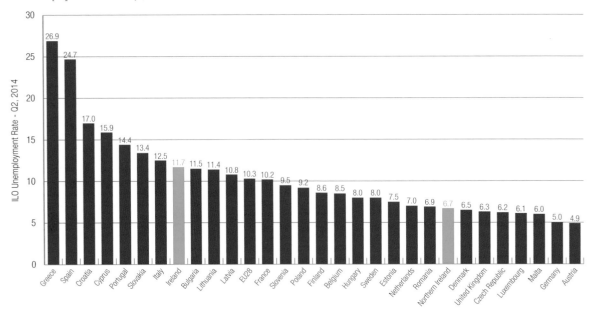

Figure 3.5: Long Term Unemployment, 2002 to 2014

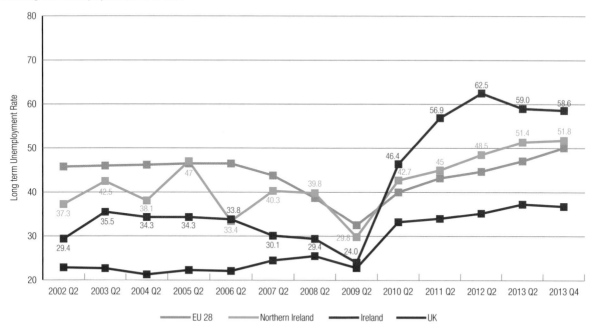

In relation to gender, there is a significant difference between unemployment rates for males and females across the island. In 2011, the unemployment rate for males in Northern Ireland was 10% with the female rate only half that at 5%. In Ireland, the male unemployment rate was 19%, compared with 12% for females. Maps 3.3 and 3.4 detail the distribution of both male and female rates across the island. For male and female unemployment rates the spatial pattern is broadly similar with highest rates in the most disadvantaged parts of our main cities, distant commuter areas and rural areas.

by an over-representation of traditional manufacturing industries. Limerick City's exceptionally high rate is partly influenced by the large amount of jobs lost in the wake of the closure of Dell manufacturing. In general, the highest levels of unemployment are observed in the peripheral West and South-East and the distant commuter zones to Dublin, which have suffered some of the worst impacts of the recession.

Among the highest levels of unemployment are to be found in our largest cities (see Map 3.2). The illustration of unemployment rates across Dublin shows that exceptionally high unemployment rates (>40%) are found in parts of the inner city, Ballyfermot and Darndale and in the 1970s new towns of Clondalkin, Ballymun and Tallaght. The lowest rates, between 5% and 10% are observed along the coast and in large parts of south county Dublin. There is also a very clear spatial pattern across the Belfast area with highest rates in the city centre and parts of West Belfast where rates are in excess of 30% in areas such as Shankill, Falls Road, Ardoyne and Twinbrook.

In Ireland, the highest male unemployment rates are recorded in Limerick City (31.9%), Waterford City (23.9%) and Mayo (23.5%), whereas highest rates in Northern Ireland are found in Derry (15.6%), Strabane (14.4%) and Limavady (14.4%). Highest female rates in Ireland are recorded in Limerick City (23.2%), North Tipperary (16.9%) and Clare (16.2%), whereas highest rates in Northern Ireland are again in the same districts of Derry (9%), Strabane (8.4%) and Limavady (7.5%).

The spatial pattern of the male unemployment rate within Dublin and Belfast City is broadly similar to that of the total unemployment rate. High rates are widespread in the north and west of the inner city and then concentrated in the peripheral disadvantaged areas of Dublin such as Darndale, Ballymun, Blanchardstown, Ballyfermot and Tallaght. In Belfast, the highest rates are in the west of the city in areas such as the Falls Road, Shankill and Turf Lodge. Female unemployment rates in our two main cities reveal a different picture with a strongly concentrated pattern of high female unemployment rates evident in Dublin's most disadvantaged areas. Belfast, on the other hand reveals a very different pattern, with rates more evenly distributed across the city centre.

MAP 3.1

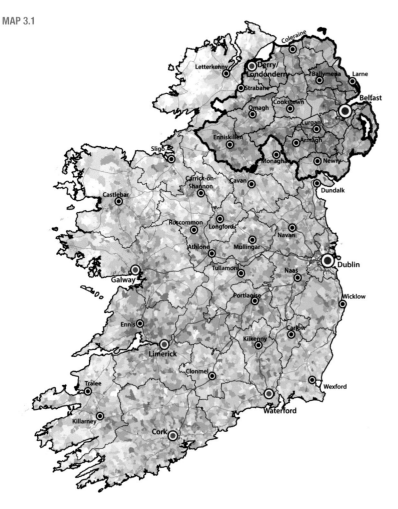

This map is part of an All-Island Atlas project developed by AIRO and the ICLRD. The project is part-financed by the European Union's INTERREG IVA programme managed by the Special EU Programmes Body.

ILO* Labour Force Participation Rate, 2011
Small Areas (SAs)
*International Labour Organisation (ILO)

Dublin City

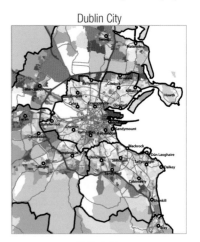

Belfast City

% ILO LF Participation Rate

- Less than 50%
- 50% to < 55%
- 55% to < 60%
- 60% to < 65%
- 65% to < 70%
- 70% to < 80%
- Greater than 80%

Northern Ireland
Local Authorities
Water

Motorways
Trunk/Primary Roads
Secondary Roads
Streets

MAP 3.2

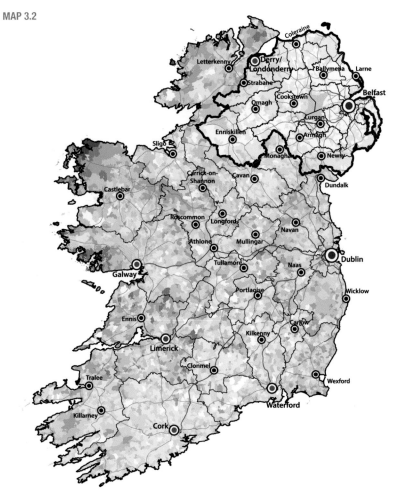

This map is part of an All-Island Atlas project developed by AIRO and the ICLRD. The project is part-financed by the European Union's INTERREG IVA programme managed by the Special EU Programmes Body.

ILO* Unemployment Rate, 2011
Small Areas (SAs)
*International Labour Organisation (ILO)

Dublin City

Belfast City

% ILO Unemployment Rate

	Less than 5%
	5% to < 10%
	10% to < 15%
	15% to < 20%
	20% to < 30%
	30% to < 40%
	Greater than 40%

	Northern Ireland
	Local Authorities
	Water

	Motorways
	Trunk/Primary Roads
	Secondary Roads
	Streets

Ordnance Survey
Ireland/Government of Ireland
Copyright Permit No. MP 005814

Crown Copyright 2014
Ordnance Survey of Northern Ireland
Permit No. 140029

Data Source: Central Statistics Office
(CSO), Northern Ireland Statistics and
Research Agency (NIRSA)

MAP 3.3

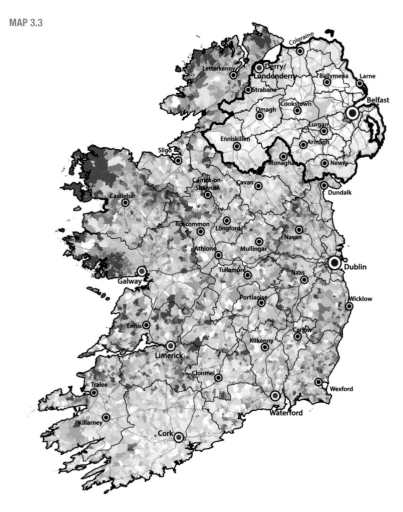

This map is part of an All-Island Atlas project developed by AIRO and the ICLRD. The project is part-financed by the European Union's INTERREG IVA programme managed by the Special EU Programmes Body.

ILO* Male Unemployment Rate, 2011
Small Areas (SAs)
*International Labour Organisation (ILO)

Dublin City

Belfast City

% ILO Unemployment Rate Male

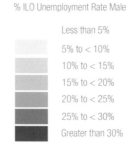

Less than 5%
5% to < 10%
10% to < 15%
15% to < 20%
20% to < 25%
25% to < 30%
Greater than 30%

Northern Ireland
Local Authorities
Water

Motorways
Trunk/Primary Roads
Secondary Roads
Streets

Ordnance Survey
Ireland/Government of Ireland
Copyright Permit No. MP 005814

Crown Copyright 2014
Ordnance Survey of Northern Ireland
Permit No. 140029

Data Source: Central Statistics Office
(CSO), Northern Ireland Statistics and
Research Agency (NIRSA)

MAP 3.4

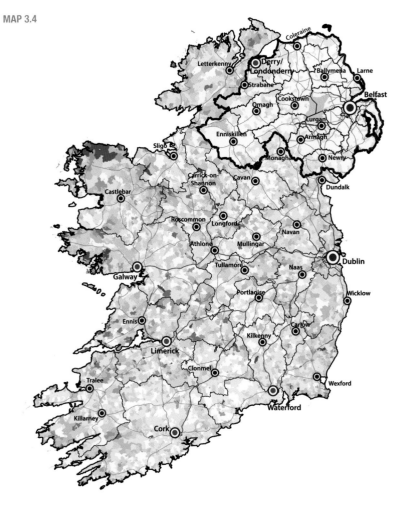

This map is part of an All-Island Atlas project developed by AIRO and the ICLRD. The project is part-financed by the European Union's INTERREG IVA programme managed by the Special EU Programmes Body.

ILO* Female Unemployment Rate, 2011
Small Areas (SAs)
*International Labour Organisation (ILO)

Dublin City

Belfast City

% ILO Unemployment Rate Female

Less than 5%
5% to < 10%
10% to < 15%
15% to < 20%
20% to < 30%
30% to < 40%
Greater than 88%

Northern Ireland
Local Authorities
Water

Motorways
Trunk/Primary Roads
Secondary Roads
Streets

Chapter 4

Industry of Employment

Dr Chris Van Egeraat

4.1 Introduction

This chapter explores the geography of *industry of employment*. For the purposes of the Census, the term *industry* is not confined to manufacturing industry. It is synonymous with the term "sector of economic activity". In this chapter, these economic activities are aggregated into seventeen broad economic groupings, with the set of maps presented below detailing the share of persons at work accounted for by these specific economic groupings. The maps presented in this chapter, therefore, reflect the level of dependence of an area on a particular industry.

The Census data relate to where the workers live, as opposed to where they conduct their work. The extent of inter-area commuting (see Chapter on Transport) means that there is a significant difference between the geography of the workers and the geography of the jobs, an issue that will become apparent in the accompanying discussion of the maps.

The geography of industry of employment reflects the legacy of the past. Over time we have witnessed the rise of new forms of economic activity, bringing new industries of employment to the fore, while the relative importance of some older industries has declined. To different extents, the evolving location patterns of industries have left an imprint on the modern industrial landscape. An interpretation of the maps is therefore facilitated by an understanding of the history of economic activity on this Island. This chapter will therefore begin with a brief history of economic activity and its geography. The subsequent section will then discuss the geography of the individual sectors.

4.2 Layers of History

Economic and industrial development of Ireland has been strongly influenced by its political and economic status within the British Empire in the 18th and 19th century. Trade restriction during the first phase of the industrial revolution meant that Ireland had not developed an internationally competitive wool and cotton manufacturing industry. The gradual dismantling of the import duties following the Act of the Union saw the closure of many mills that were unable to withstand the competition from more efficient producers in England and Scotland. This stands in stark contrast to the fortunes of linen manufacturing. The inventions that supported the mechanisation of linen spinning only occurred in the 1820s and entrepreneurs in Ulster, the traditional centre of the linen cottage industry, developed a competitive linen manufacturing centre, increasingly focused on Belfast. The demand for machinery stimulated the engineering industry which in turn attracted shipbuilding, developing Belfast into a major manufacturing centre.

Dublin had developed into a typical colonial capital, disproportionate in size to other urban centres, and the centre of administration and transport. It therefore had a disproportionate share of employment in Public Administration; Education; Professional, Scientific & Technical as well as Transport & Storage. The rest of the Island, with the exception of the North East, had a lack of significant urban centres. It became increasingly focused on agriculture with a disproportionately low share of administrative and manufacturing employment.

By the time of the formation of the free state, the two parts of the Island had developed in different ways. Ireland, having lost its main industrial centre, was now a strongly agricultural economy. Over half of the population was employed in agriculture. Manufacturing industry, mainly concentrated in Dublin, accounted for a mere four percent of employment. In 1932 the government adopted a strategy of import substitution industrialisation in order to stimulate indigenous manufacturing. Tariff barriers were erected and foreign ownership of manufacturing discouraged. The policy led to strong growth in manufacturing employment, mainly in the East of the country. In the West, however, manufacturing employment actually fell. The initial gains of this policy would be short lived. The indigenous industry was uncompetitive and not successful in export markets, where it faced retaliatory import duties. The industry mainly focused on the "low-hanging fruit": the small size of the home market and lack of technological knowledge meant that more sophisticated manufacturing products and capital goods had to be imported. The resultant crisis in the balance of payments necessitated a change in economic policy.

This policy change came towards the end of the 1950s in the form of an export-oriented economic policy characterised by a gradual dismantling of the tariff barriers, fiscal export incentives and the attraction of foreign direct investment. Indigenous manufacturing suffered a setback, but, due to the influx of foreign companies, overall manufacturing employment grew strongly in the 1960s and 1970s. A disproportionate share of the employment in foreign manufacturing companies occurred outside Dublin, partly driven by the locational preferences of the companies involved and partly by the active policy of the Industrial Development Authority to spread industrial employment in proportion to the distribution of the population. A disproportionate share of the new jobs ended up in locations away from the main urban centres, while Dublin experienced a disproportionate decline in employment in indigenous manufacturing companies. The limited resources available for high-spec infrastructural development mean that these resources were concentrated in a small number of locations, which eventually led to a chemical processing cluster in Cork.

Meanwhile the process of tertiarisation of the Irish economy also began to emerge in these decades. In 1950, the services sector accounted for 36% of employment, an increase of two percentage points since 1926. But this share grew steadily, driven by the growth of government and corporate bureaucracy, a growing demand for education, a growing demand for business services and, of course, the mechanisation of agriculture. By 1981, about half of the workers were employed in services and by 1991 this figure has risen to 59%. The Celtic Tiger heralded a significant restructuring of the Irish Economy. Foreign investment in manufacturing continued but had become increasingly concentrated in more sophisticated activities such as pharmaceuticals, medical devices and electronics. After 2000, employment in low value-added segments of electronics manufacturing began to fall. Due to its high value-added nature, new manufacturing employment in foreign firms became increasingly focussed on the main urban areas such as Dublin, Cork and Galway. Another important development was the growing share of services in the foreign segment. Internationally-traded services such as Software and Financial Services began to account for the majority of new foreign jobs. This rapidly growing sector became strongly focused on Dublin.

The overall process of tertiarisation continued such that by 2006, 71% of persons were employed in services. The main growth areas included Other Business Services (including internationally-traded services), Health & Social Work, Finance & Insurance and Hotels & Restaurants. Another notable development was the growth in construction employment on the back of the construction boom. Between 1991 and 2006 construction employment grew over 180% to 215,000 employees.

The global financial and economic crisis which emerged in 2007 caused serious unemployment, but also a significant change to the industrial structure. The implosion in construction caused a 58% fall in construction employment between 2006 and 2011, reaching an unnatural low of 5% of total employment. Probably more permanent is the 21% fall in manufacturing employment, representing a drop of two percentage points of total employment.

Northern Ireland's experience since partition has been very different. The economy was initially much stronger than that of Ireland, dominated by agriculture and a well developed indigenous manufacturing sector. Shipbuilding, heavy engineering, and textiles were concentrated in the East of the country. Services were boosted by Belfast's new role as a regional administrative centre. When, during the 1950s, some of the traditional manufacturing sectors started to shed employment, the region was initially successful in attracting foreign

investment. However, the onset of political unrest in the 1970s put a halt to this. At the same time, shipbuilding and other heavy industry, facing global competition, went into decline. The fall in agricultural and manufacturing employment meant that during the 1980s and 1990s, the Northern Ireland economy became increasingly dependent on public services, partly supported by subventions. Construction sector employment in Northern Ireland has proved to be more stable than in Ireland.

4.3 Industry in Ireland in 2011
The Island of Ireland in 2011 should be characterised as services or informational economy. The agricultural sector, once the main employer, is now accounting for a very small share of the workers (4.2%). The share of manufacturing, although higher than in some other EU countries, is small (10%) and declining rapidly. The vast majority of workers are now employed in services activities, including the high-wage informational economy services.

However, the spatial distribution of the various activities differs from industry to industry, partly reflecting the developments outlined above. At the highest level of spatial aggregation we observe significant differences between Ireland and Northern Ireland (see Figure 4.1). The most important dissimilarities relate to Wholesale and Retail, Human Health and Social Work and Construction activities, which account for a far greater share of employment in Northern Ireland compared to Ireland. In Ireland, Agriculture & Forestry and the relatively high-wage Financial and Insurance activities are accounting for a relatively high share of workers. But there are also important differences between the two jurisdictions. Maps 4.1 - 4.11 present the importance of selection of different industrial groupings at the Small Area (SA) level across the Island in 2011.

Map 4.1 illustrates that, although accounting for a small share of employment on this Island, Agriculture, Forestry and Fishing remains a very important industry in many areas. This is, of course, driven by the fact that the sector is an important employer in the many sparsely populated SAs that cover an extensive area of the Island while, in contrast, it is a relatively insignificant employer in the more populous towns and cities. The sector is clearly of greater importance in Ireland than in Northern Ireland. In Ireland it accounts for over 20% of employment in many SAs. The traditional East-West divide, which goes back to colonial times, is still discernible, both in Ireland and in Northern Ireland, with large areas of Kerry, West-Cork

and Mayo heavily dependent (>30%) on Agriculture. At a local authority level, the industry is most important in Cavan (12.3%) and Monaghan (12.2%), partly driven by the lack of significant urban areas. In Northern Ireland, the peripheral Moyle District is most dependent on Agriculture (6.5%)

Interpreting the geography of employment in Manufacturing (Map 4.2) is a less straightforward task. The traditional importance of indigenous industry for the Belfast area and Dublin is no longer discernible. Much of this industry has disappeared in these areas and is overshadowed by the growth of services. In Ireland, the manufacturing sector has remained important in other areas for various reasons. The active policy of dispersal of foreign investment in manufacturing during the 1960s and 1970s is reflected in the fact that the sector remains relatively important in many SAs outside the Dublin area, including in the main urban centres of Waterford City (16.4%), Cork City (12.4%) and Galway City (13.2%). Commuting partly explains the importance of the sector in the SAs near these urban centres. The high levels of employment to the west of Killybegs, Co. Donegal is probably indicative of the seafood processing activity located in what is a sparsely populated area. In Northern Ireland manufacturing has remained a relatively important industry in Dungannon (19.5% - probably picking up those employed in the large food processing plants), Craigavon (16.3%) and Cookstown (16.1%).

Mining and Quarrying employment (Map 4.3) is obviously directly linked to the location of natural resources. The peat harvesting and milling activities of *Bord na Mona* are clearly picked up in the areas to the west and the east of Tullamore town, providing employment in a sparsely populated area. The same holds for the base-metal mining activities at the Lisheen, Galmoy (west of Kilkenny town) and Navan deposits.

Turning to services activities, a number of these activities are located in many towns. This, in combination with inter-SA commuting paints a kaleidoscopic picture lacking strong patterns. Good examples of this include Wholesale and Retail (Map 4.4), Human Health and Social Work (Map 4.5) and Education (Map 4.6). Retail activity tends to be concentrated in urban centres but is an important employer for workers living in many areas. It is a particularly important sector in Northern Ireland, where it accounts for between 15% and 20% of employment in nearly all Districts. Interesting is the doughnut pattern in the Dublin area, with Dublin City

(11.7%) and Dún Laoghaire-Rathdown (12.3%) among the bottom three local authorities in the country. This is likely a reflection of the lower salary levels in the sector, which has pushed the workers out of the Dublin housing market and into the outer suburbs. Human Health and Social work is also relatively important in Northern Ireland, accounting for between 13% and 17% of employment in nearly all Districts. In Ireland, the most notable feature is the relative importance of the sector to Sligo (14.7%) and Leitrim (13%), which is most likely driven by a deficit in employment opportunities in other sectors.

Construction activity (Map 4.7) presents a similarly kaleidoscopic picture, though there are some notable features. Firstly, the industry is distinctly more important to areas in Northern Ireland than to areas in Ireland. To an extent this reflects the impact of the collapse in the construction industry, which was far more severe in Ireland. However, the sharpness of the divide between

SAs along the border suggests that classification issues may play a role here. The relatively limited role of the sector to Dublin is also apparent with only 2.8% of the workers in Dublin City employed in this sector, the lowest figure on the Island.

The distribution of the importance of employment in Public Administration, Defence and Security displays a more definite pattern (Map 4.8). In Ireland, a number of areas in the Midlands, West and North-West are characterised by a relatively high dependence on Public Administration, in many cases reflecting a deficit in other economic activities. Sligo is the county with the highest dependence (8%) on Public Administration. Although a large share of Public Administration jobs is concentrated in Dublin, the industry makes up a more modest share of Dublin's total employment (6.4%). As discussed, partly supported by subvention, the industry is relatively prominent in Northern Ireland. The areas most dependent on the industry are concentrated in and around Belfast.

The industry accounts for over 10% of employment in the districts North Down, Castlereagh, Ards and Lisburn. A number of services activities are concentrated in metropolitan areas and these sectors are therefore of particular importance to these cities and their commuting belts. One example is Financial and Insurance activities (Map 4.9). From colonial times, Dublin has acted as the domestic banking and insurance centre. Its position was strengthened by the establishment of the International Financial Services Centre at the end of the 1980s. Unlike the situation described in relation to retail activities, many workers in this sector are better able to afford housing in Dublin. The top four in terms of dependence encompass the four Dublin local authorities. The industry is of particularly important to Dún Laoghaire-Rathdown, where it accounts for 12.5% of workers. That said, quite a number of workers are pushed into the commuting belt, with Wicklow, Kildare and Meath ranking immediately below the Dublin regions in terms of dependence on Financial and Insurance activities.

Figure 4.1: Industry of Employment Ireland and Northern Ireland, 2011

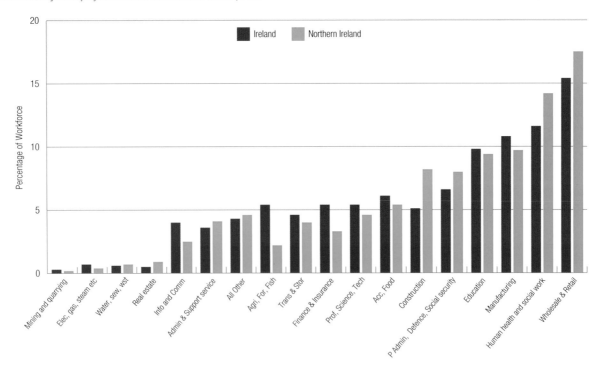

In Northern Ireland, the sector is again relatively more important to Belfast and surrounding districts but here the differences are less pronounced due to the smaller size of the sector. Professional, Scientific and Technical employment (Map 4.10) is similarly concentrated in Dublin and Belfast. The top five in terms of dependence is made up of the four Dublin authorities and Wicklow, with Dún Laoghaire-Rathdown again having the highest

dependence (10.7%). In Belfast, the industry accounts for 6.2% of employment.

Finally, Transport and Storage activities (Map 4.11) are clearly reflecting the location of the transport and logistics infrastructure. SAs near large harbour facilities and airports are for obvious reasons more dependent on this industry and can easily be picked up from the map.

In Northern Ireland, Larne is the District with the highest share of workers employed in this sector (7.5%). In the Ireland we find relatively high dependencies in Fingal (8.2%) South Dublin (6%) and Meath (6%), driven by Dublin Airport and the logistics centres along the M50 Motorway.

MAP 4.1

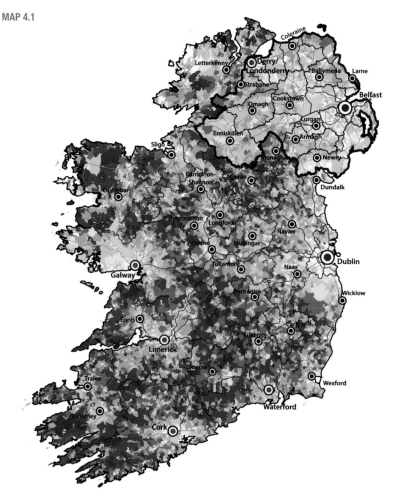

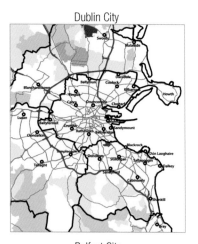

Percentage of Workforce Employed in:
Agriculture, Forestry and Fishing
Small Areas (SAs)

Dublin City

Belfast City

This map is part of an All-Island Atlas project developed by AIRO and the ICLRD. The project is
part-financed by the European Union's INTERREG IVA programme managed by the Special EU Programmes Body.

% IoE: Agri, For & Fishing

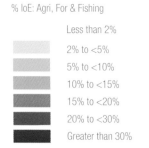

Less than 2%
2% to <5%
5% to <10%
10% to <15%
15% to <20%
20% to <30%
Greater than 30%

Northern Ireland
Local Authorities
Water

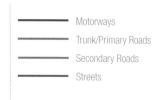

Motorways
Trunk/Primary Roads
Secondary Roads
Streets

Ordnance Survey
Ireland/Government of Ireland
Copyright Permit No. MP 005814

Crown Copyright 2014
Ordnance Survey of Northern Ireland
Permit No. 140029

Data Source: Central Statistics Office
(CSO), Northern Ireland Statistics and
Research Agency (NIRSA)

MAP 4.2

Percentage of Workforce Employed in:
Manufacturing
Small Areas (SAs)

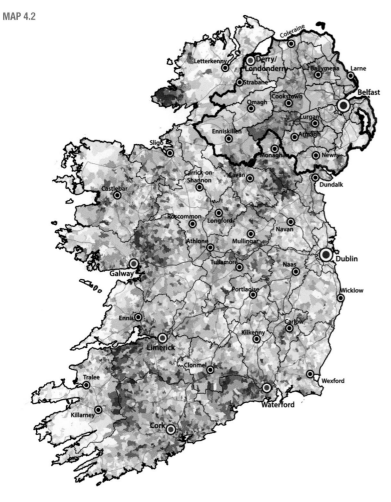

Dublin City

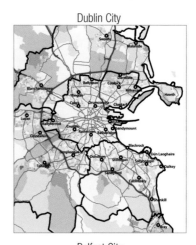

Belfast City

This map is part of an All-Island Atlas project developed by AIRO and the ICLRD. The project is part-financed by the European Union's INTERREG IVA programme managed by the Special EU Programmes Body.

% IoE: Manufacturing

Less than 5%
5% to <8%
8% to <12%
12% to <16%
16% to <20%
20% to <28%
Greater than 28%

Northern Ireland
Local Authorities
Water

Motorways
Trunk/Primary Roads
Secondary Roads
Streets

Ordnance Survey
Ireland/Government of Ireland
Copyright Permit No. MP 005814

Crown Copyright 2014
Ordnance Survey of Northern Ireland
Permit No. 140029

Data Source: Central Statistics Office
(CSO), Northern Ireland Statistics and
Research Agency (NIRSA)

MAP 4.3

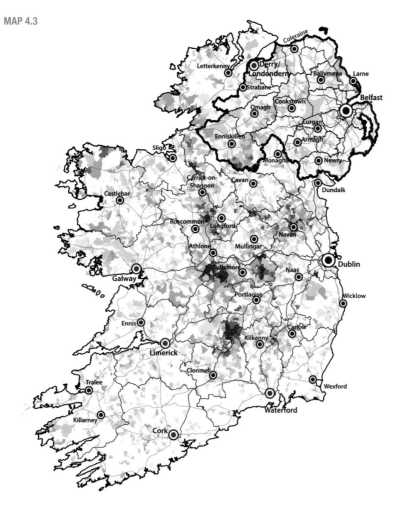

This map is part of an All-Island Atlas project developed by AIRO and the ICLRD. The project is
part-financed by the European Union's INTERREG IVA programme managed by the Special EU Programmes Body.

Percentage of Workforce Employed in:
Mining and Quarrying
Small Areas (SAs)

Dublin City

Belfast City

% IoE: Mining and Quarrying

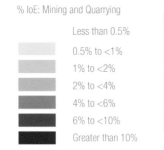

	Less than 0.5%
	0.5% to <1%
	1% to <2%
	2% to <4%
	4% to <6%
	6% to <10%
	Greater than 10%

Northern Ireland
Local Authorities
Water

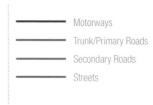

Motorways
Trunk/Primary Roads
Secondary Roads
Streets

Ordnance Survey
Ireland/Government of Ireland
Copyright Permit No. MP 005814

Crown Copyright 2014
Ordnance Survey of Northern Ireland
Permit No. 140029

Data Source: Central Statistics Office
(CSO), Northern Ireland Statistics and
Research Agency (NIRSA)

MAP 4.4

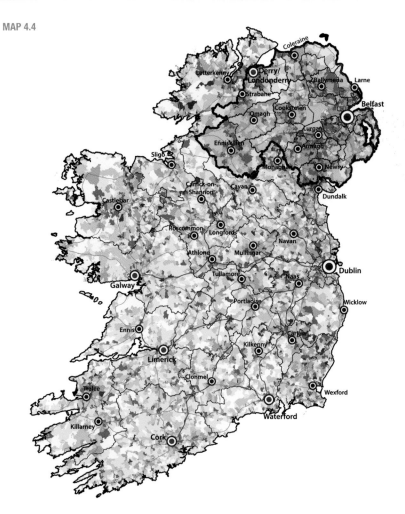

Percentage of Workforce Employed in:
Wholesale and Retail*
Small Areas (SAs)
*Wholesale and retail trade, repair of motor vehicles and motorcycles

Dublin City

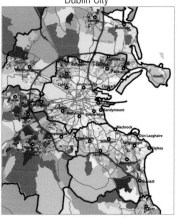

Belfast City

% IoE: Wholesale & Retail

	Less than 8%
	8% to <12%
	12% to <15%
	12% to <16%
	16% to <20%
	20% to <28%
	Greater than 28%

	Northern Ireland
	Local Authorities
	Water

	Motorways
	Trunk/Primary Roads
	Secondary Roads
	Streets

MAP 4.5

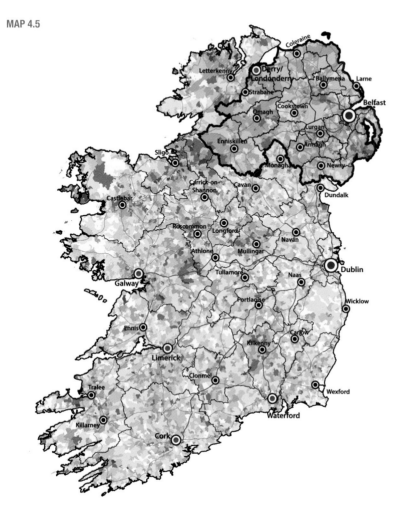

Percentage of Workforce Employed in:
Human Health and Social
Small Areas (SAs)

Dublin City

Belfast City

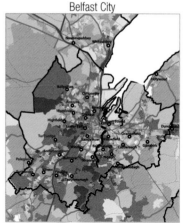

This map is part of an All-Island Atlas project developed by AIRO and the ICLRD. The project is part-financed by the European Union's INTERREG IVA programme managed by the Special EU Programmes Body.

% IoE: Health and Social

	Less than 6%
	6% to <10%
	10% to <13%
	13% to <17%
	17% to <22%
	22% to <32%
	Greater than 32%

	Northern Ireland
	Local Authorities
	Water

	Motorways
	Trunk/Primary Roads
	Secondary Roads
	Streets

Ordnance Survey
Ireland/Government of Ireland
Copyright Permit No. MP 005814

Crown Copyright 2014
Ordnance Survey of Northern Ireland
Permit No. 140029

Data Source: Central Statistics Office
(CSO), Northern Ireland Statistics and
Research Agency (NIRSA)

MAP 4.6

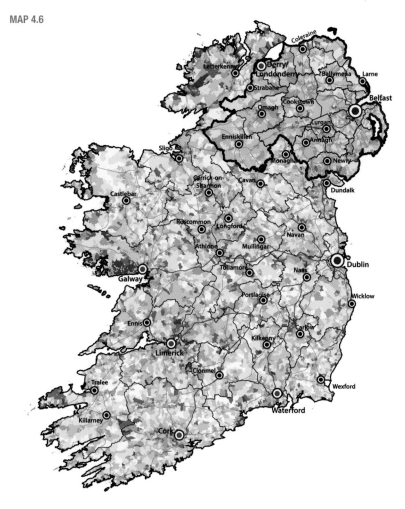

Percentage of Workforce Employed in:
Education
Small Areas (SAs)

Dublin City

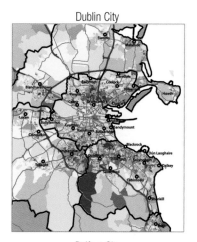

Belfast City

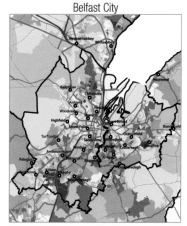

This map is part of an All-Island Atlas project developed by AIRO and the ICLRD. The project is part-financed by the European Union's INTERREG IVA programme managed by the Special EU Programmes Body.

% IoE: Education

Less than 5%
5% to <8%
8% to <11%
12% to <14%
14% to <18%
18% to <36%
Greater than 36%

Northern Ireland
Local Authorities
Water

Motorways
Trunk/Primary Roads
Secondary Roads
Streets

Ordnance Survey
Ireland/Government of Ireland
Copyright Permit No. MP 005814

Crown Copyright 2014
Ordnance Survey of Northern Ireland
Permit No. 140029

Data Source: Central Statistics Office
(CSO), Northern Ireland Statistics and
Research Agency (NIRSA)

MAP 4.7

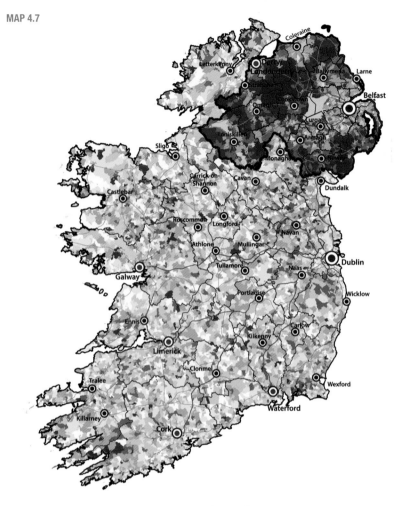

Percentage of Workforce Employed in:
Construction
Small Areas (SAs)

Dublin City

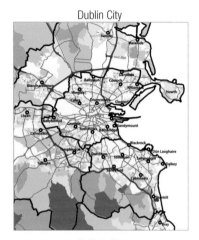

Belfast City

This map is part of an All-Island Atlas project developed by AIRO and the ICLRD. The project is part-financed by the European Union's INTERREG IVA programme managed by the Special EU Programmes Body.

% IoE: Construction

- Less than 3%
- 3% to <5%
- 5% to <7%
- 7% to <10%
- 10% to <13%
- 13% to <18%
- Greater than 18%

- Northern Ireland
- Local Authorities
- Water

- Motorways
- Trunk/Primary Roads
- Secondary Roads
- Streets

MAP 4.8

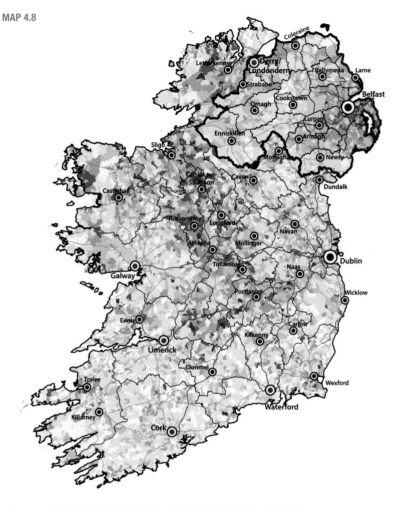

This map is part of an All-Island Atlas project developed by AIRO and the ICLRD. The project is part-financed by the European Union's INTERREG IVA programme managed by the Special EU Programmes Body.

Percentage of Workforce Employed in:
Public Admin, Defence and Security
Small Areas (SAs)

Dublin City

Belfast City

% IoE: Public Admin & Defence

- Less than 3%
- 3% to <6%
- 6% to <8%
- 8% to <11%
- 11% to <15%
- 15% to <33%
- Greater than 33%

Northern Ireland
Local Authorities
Water

Motorways
Trunk/Primary Roads
Secondary Roads
Streets

Ordnance Survey
Ireland/Government of Ireland
Copyright Permit No. MP 005814

Crown Copyright 2014
Ordnance Survey of Northern Ireland
Permit No. 140029

Data Source: Central Statistics Office
(CSO), Northern Ireland Statistics and
Research Agency (NIRSA)

MAP 4.9

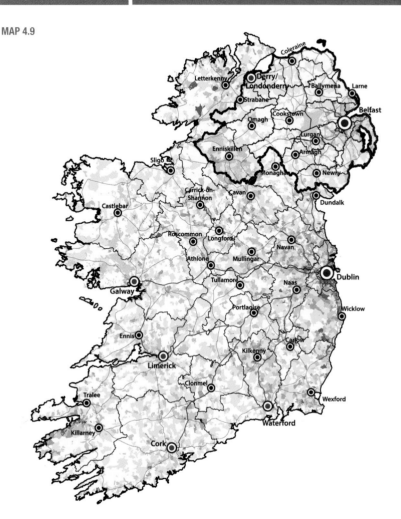

Percentage of Workforce Employed in:
Financial and Insurance
Small Areas (SAs)

Dublin City

Belfast City

This map is part of an All-Island Atlas project developed by AIRO and the ICLRD. The project is part-financed by the European Union's INTERREG IVA programme managed by the Special EU Programmes Body.

% IoE: Finance & Insurance

Less than 2%
2% to <4%
4% to <6%
6% to <9%
9% to <13%
13% to <17%
Greater than 17%

Northern Ireland
Local Authorities
Water

Motorways
Trunk/Primary Roads
Secondary Roads
Streets

Ordnance Survey Ireland/Government of Ireland Copyright Permit No. MP 005814

Crown Copyright 2014 Ordnance Survey of Northern Ireland Permit No. 140029

Data Source: Central Statistics Office (CSO), Northern Ireland Statistics and Research Agency (NIRSA)

MAP 4.10

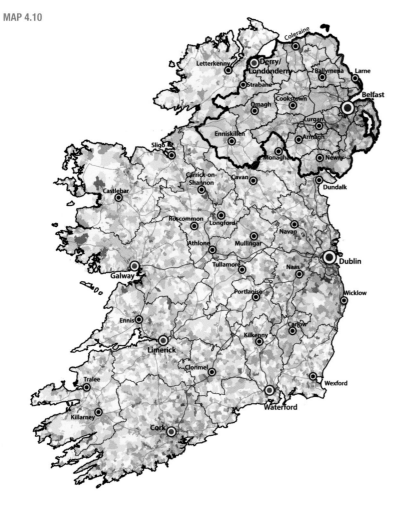

This map is part of an All-Island Atlas project developed by AIRO and the ICLRD. The project is
part-financed by the European Union's INTERREG IVA programme managed by the Special EU Programmes Body.

Percentage of Workforce Employed in:
Professional, Scientific & Technical
Small Areas (SAs)

Dublin City

Belfast City

Legend

% IoE: Prof, Sci & Technical

- Less than 2%
- 2% to <4%
- 4% to <6%
- 6% to <9%
- 9% to <13%
- 13% to <18%
- Greater than 18%

- Northern Ireland
- Local Authorities
- Water

- Motorways
- Trunk/Primary Roads
- Secondary Roads
- Streets

Ordnance Survey
Ireland/Government of Ireland
Copyright Permit No. MP 005814

Crown Copyright 2014
Ordnance Survey of Northern Ireland
Permit No. 140029

Data Source: Central Statistics Office
(CSO), Northern Ireland Statistics and
Research Agency (NIRSA)

MAP 4.11

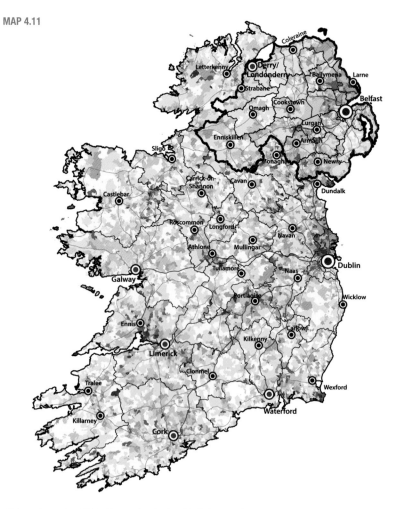

Percentage of Workforce Employed in:
Transport and Storage
Small Areas (SAs)

Dublin City

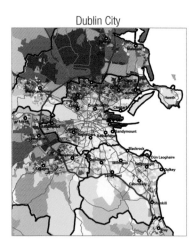

Belfast City

This map is part of an All-Island Atlas project developed by AIRO and the ICLRD. The project is
part-financed by the European Union's INTERREG IVA programme managed by the Special EU Programmes Body.

% IoE: Transport & Storage

Less than 1.5%
1.5% to <3.5%
3.5% to <5%
5% to <7.5%
7.5% to <10%
10% to <14%
Greater than 14%

Northern Ireland
Local Authorities
Water

Motorways
Trunk/Primary Roads
Secondary Roads
Streets

Chapter 5

Education

Prof. James A Walsh

5.1 Introduction

The highest level of education completed is probably the single most important attribute of the population. The progressive improvements in formal education provision and increasing levels and duration of participation by successive cohorts have been major contributors to the economic, social, cultural and political transformations that have occurred over several decades. Enhanced levels of education among increasing proportions of successive age cohorts have contributed significantly to economic growth through higher participation rates in the workforce, increased productivity as more workers have the knowledge and skills to obtain employment in industrial sectors with high levels of added value, and greater levels of disposable income to support increasing expenditure on personal consumption items. Higher education levels also contribute greatly to social change through, for example, lower fertility rates, increased female participation in the paid workforce, improvements in health leading to longer life expectancy, and higher levels of social responsibility. On a more general level improvements in education have also been associated with the emergence of a more assertive, critical, open and internationally engaged society that is well prepared to adapt to and benefit from the opportunities associated with new technologies and increased globalisation.

The education geographies illustrated on Maps 5.1 to 5.5 are the outcomes recorded in 2011 from the interaction over many years of a number of inter-related factors. These include place specific demographic influences such as age and gender profiles, and histories of inter-county and international migrations, both inwards and outwards. Also important are the influences on the location patterns of different types of employment opportunities, especially the location of public services such as education, health and public administration, the rapidly growing new professional areas, different types of manufacturing, and the more traditional natural resource based industries. These location influences are particularly important for shaping the role of gender in the education geographies of the total population. The location of tertiary education opportunities is becoming of greater significance as participation levels increase and more opt to attend the nearest higher education institution providing their preferred programmes of study.

A further influence especially in the larger urban centres is the extent of spatial segregation of different social classes through public housing programmes (See Chapter 7 on Housing). Place-specific and related social status factors also impact on the dynamic of inter-generational shifts in education attainment levels and their role in reinforcing or disrupting spatial patterns. For example, a twenty year old member of a household where both parents have a low level of education has a likelihood of between 33% and 45% of still being involved in education. For households where both parents have completed their education at upper secondary level the likelihood rises to 68%, and further to 92% and 94% where both parents have either an honours bachelor degree or a postgraduate qualification.

Following the introduction of 'free' post-primary education in Ireland in the late 1960s and the massification of higher education in the 1990s there has been a very significant increase in the age at which full-time education ceases for the majority of young people. In 1971 over three-fifths (62.4%) of the those aged 15 years and over who had already ended their full-time education, had left the education system before reaching the age of 16. By 1981 the proportion had dropped to 50% and it declined further to 40% by 1991. In 2002 the proportion was just under one-quarter and by 2011 it was only 13%. Conversely, the proportion for whom full-time education ceased after reaching the age of 19 increased from 6.6% in 1971 to almost 30% in 2011. Despite the general decline in the number of persons completing their formal education before the age of sixteen there were still 38,257 aged 15-24 in 2011 in Ireland who had completed their education at either primary or lower secondary levels. This subset amounting to 6.6% of the 15-24 age cohort are generally very poorly prepared for participation in the labour force and are at risk of experiencing unemployment and related social challenges throughout their lives.

The increasing duration of full-time education experience is reflected in the highest level of education completed. The proportion of persons who had left the education system with only primary level or no formal education declined from 64% in 1971 to 20% in 2011. Over the same period the proportion that completed their education with a third level qualification increased from four to 29%. (Table 5.1)

Table 5.1: Percentage distribution of the population in Ireland whose full-time education has ceased classified by highest level of education completed in Ireland, 1971-2011

Year	Highest level of education completed		
	Primary*	Second level	Third level
1971	63.7	32.0	4.3
1981	45.5	46.8	7.7
1991	36.7	50.2	13.1
1996	31.8	49.1	19.0
2002	26.2	49.2	24.7
2011	19.9	51.0	29.1

* including no formal education and not stated

The trend towards progressively higher educational attainment levels is illustrated for selected age cohorts in Table 5.2.

Table 5.2: Percentage distribution aged 15 years and over, whose education has ceased, by highest level of education completed and age group in Ireland, 2011

Age Group	Lower Secondary or Less	Upper Secondary	Third Level	Not Stated	Total
15 - 24	19.9	53.3	22.1	4.6	100.0
25 - 44	16.9	38.0	41.5	3.6	100.0
45 - 64	39.3	33.2	24.0	3.5	100.0
65 +	58.6	20.8	11.8	8.8	100.0
Total	31.9	34.4	29.1	4.7	100.0

In 2011 just under three fifths (58.6%) of the population aged 65 years and over had only completed lower secondary education or less. The majority had only completed primary education. A decade earlier the proportion was 65%, and within another decade it is likely to decline to about 47%. In the cohort aged 25 - 44 in 2011 only 17% have completed their formal education at lower secondary level or less. By contrast, the proportions that have completed their education with a third level award increased from 12% for those aged 65 years and over to 41.5% for those aged 25-44 years. The comparable proportions one decade earlier were only 9% and 35% respectively. The lower third level proportion for those aged 15-24 is simply due to the fact that 65% of that specific cohort are still in education either as undergraduates or postgraduates.

In the Northern Ireland census the data on education is collated for persons aged 16 years and over. The major trends noted above are also evident in Northern Ireland. The most notable difference is that participation in post-junior level education increased earlier and this was subsequently followed by higher proportions progressing to third level some decades prior to the same trend emerging in Ireland. Thus, higher proportions of the cohorts aged 45 years and over in Northern Ireland have a third level qualification whereas the reverse is the case for those aged under forty years in 2011. The transition in third level participation rates can be linked to the late 1980s.

In order to map for the first time education levels for the whole island of Ireland the Northern Ireland Statistics and Research Agency in cooperation with the Central Statistics Office have produced a database for the population aged 16 years and over in 2011 consisting of five categories that are summarised in Table 5.3.

5.2 Low Education

The geographical distribution of the population aged 16 years or more who had completed their education at either lower secondary level or less, or at either level 1 or without a qualification in Northern Ireland displays a marked contrast between high proportions in some of the most rural areas in the west and border counties and much lower proportions in the principal city regions (Map 5.1). There is a significant contrast in levels of education completed between the populations of the 'aggregate urban' and 'aggregate rural' areas as defined by the Central Statistics Office. In 2011 in the aggregate rural areas 36.5% of the population aged 15 years and over had completed their education at a low level compared to 28.9% for the aggregate urban population. The significant progress between 2002 and 2011 is evident from the reduction in the rural and urban proportions from 49.5% and 37.9% respectively, with the greater adjustment in rural areas.

The individuals that have completed their education at a low lever account for the largest shares of the population: 37% in Ireland and 41% in Northern Ireland. Over 50% of the populations are included in this category in parts of Donegal, west Mayo, west Galway, southwest Clare, the southwest peninsulas, the Kerry-Limerick-Cork borderland, parts of the Midlands, extensive parts of the counties south of the Border with Northern Ireland, and within Northern Ireland in some of remotest western rural areas. These areas share characteristics of high proportions of elderly populations and long established patterns of selective net out-migration to other parts of Ireland and abroad. These movements have frequently contributed to an exodus of those who had opportunities to complete higher levels of formal education. Comparatively high percentages also occur in many traditional industrial towns such as Dundalk, Drogheda, Ardee, Duleek, Arklow, Wexford, Enniscorthy, New Ross, Longford, Thurles, Tipperary, Athy, Monasterevan, Roscrea, Edenderry, Clara, Portarlington, Mountmellick and Ballinasloe.

In Dublin there is a pronounced contrast, linked to social class distributions, between districts in the north and west and those areas with mostly higher levels of education in the south and east. In Belfast there is also a pronounced divide between central and western districts

Table 5.3: Highest level of education of usual residents aged 16 and over, 2011

Ireland			Northern Ireland		
Highest level of education completed	**No.**	**%**	**Highest level of education completed**	**No.**	**%**
None (incl. not stated), primary, lower secondary	1285800	36.8	No qualification, level 1	581649	40.6
Upper secondary and higher certificate	898932	25.7	Level 2 and 3 (e.g., 5 or more GCSEs)	389680	27.2
Apprenticeship	188473	5.4	Apprenticeship	60462	4.2
Vocational	299055	8.6	Vocational, other	61205	4.3
Third level (degree or higher)	822512	23.5	Level 4 and above (degree or higher)	338544	23.6
Total	**3494772**	**100.0**		**1431540**	**100.0**

Source: CSO and NIRSA, 2011

such as Shankill, Lower Falls, Sandy Row, Turf Lodge, High Field and Andersonstown and the more affluent middle class areas in the south and east.

5.3 Medium Education
The geographical variation in the distribution of persons that have completed a medium level of education (upper secondary and higher certificate) is less pronounced. The differential between both jurisdictions is less: 25.7% in Ireland vs. 27.2% in Northern Ireland. The proportions are less than 20% in the rural areas with the weakest economic structures and less than 24% throughout much of the remaining open countryside. Higher proportions are evident in the districts that are adjacent to many of the towns. In Northern Ireland the proportions are higher throughout more of the rural parts in the East. The most striking features of both Dublin and Belfast are the relatively low proportions of the populations with medium education levels, reflecting the polarisation between areas characterised by either low or high education levels. In south Dublin there are higher proportions with middle than low levels of education in contrast to many areas in the north and west of the city (Map 5.2).

5.4 High Education
The proportions of the populations aged 16 years and over in both jurisdictions that have completed their education with third level qualifications are identical at 23.5%. Their distribution is much more spatially concentrated with clear distinctions between the city regions and most of the remainder of the country (Map 5.3). However, the level of concentration in Dublin and the three surrounding counties declined from 48% in 2002 to 46% in 2011 and the differential between

aggregate urban and rural areas weakened. In 2011 the proportion of the rural resident population with third level qualifications at 24.4% was 7.8 percentage points less than the urban proportion at 32.1%. In 2002 the comparable proportions were 19.1% and 28.5%, a differential of 9.4 percentage points.

Some key influences on the spatial distribution of the most highly educated are the location of universities and other higher education institutions, the locations of advanced medical centres especially the specialist regional hospitals, and also the location of the public administration offices in each county. The cities of Dublin, Belfast, Derry/Londonderry, Cork, Limerick and Galway are distinguished by spatially extensive hinterlands where at least 20% of those adults for whom formal education had already ended had achieved a third level qualification. Further concentrations are also evident in the vicinity of most towns with populations greater than 10,000 persons, especially those where Institutes of Technology are located, e.g., Dundalk, Carlow, Tralee, Athlone, Sligo and Letterkenny, and also in the university town of Maynooth in north Kildare. In addition there are above average percentages with third level education in towns such as Kenmare, Dingle and Westport that have attracted high levels of immigration.

Within Dublin there is a pronounced concentration of the highly educated population in the southeast especially in Dún Laoghaire-Rathdown where 49.5% of those aged 15 years and over have a third level qualification, and also to a lesser extent in the northeast and in Castleknock in the west. In Belfast the highly educated population is also concentrated around the university and the east.

5.5 Vocational and Apprenticeship
Maps 5.4 and 5.5 show distributions of the populations that completed their education with either an apprenticeship or a vocational award. The apprenticeship courses are typically placed on level 6 of the National Framework of Qualifications and tend to be provided by agencies supporting manufacturing, the marine sector, agriculture, crafts, and also catering and culinary activities. The vocational qualifications tend to be in similar areas but placed at lower levels on the national framework.

There are more than three times as many with an apprenticeship award in Ireland as there are in Northern Ireland where they account for 5.4% and 4.2% of the populations aged 16 years and over. The importance of apprenticeships is particularly evident in the Midlands and Southeast, and also in parts of the Southwest and Mid-West. They are much less prevalent in the West and Northwest and in the Border counties apart from county Cavan. In Northern Ireland they are almost absent from much of the western and southern districts. In Dublin and Belfast the highest concentrations are in some of the more working class outer suburban districts.

There are nearly 300,000 persons with vocational qualification in Ireland compared to just over 61,000 in Northern Ireland. They are widely dispersed especially throughout rural areas, though again they are comparatively less prevalent in the West, Mid-West and Southwest. In Dublin they are mostly concentrated in a necklace of districts located beyond the M50.

MAP 5.1

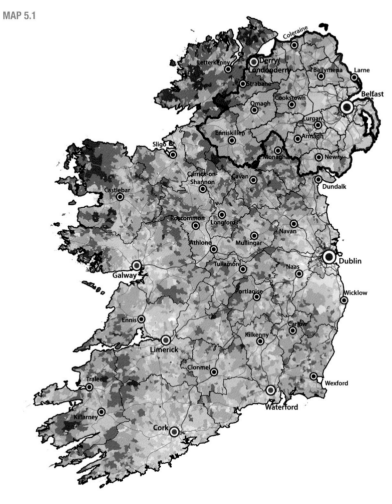

This map is part of an All-Island Atlas project developed by AIRO and the ICLRD. The project is part-financed by the European Union's INTERREG IVA programme managed by the Special EU Programmes Body.

Highest level of education: Low*
Small Areas (SAs)
* None (incl. not stated, primary & lower secondary)

Dublin City

Belfast City

% Education: Low

	Less than 19%
	19% to < 30%
	30% to < 39%
	39% to < 48%
	48% to < 60%
	Greater than 60%

Northern Ireland
Local Authorities
Water

Motorways
Trunk/Primary Roads
Secondary Roads
Streets

Ordnance Survey
Ireland/Government of Ireland
Copyright Permit No. MP 005814

Crown Copyright 2014
Ordnance Survey of Northern Ireland
Permit No. 140029

Data Source: Central Statistics Office
(CSO), Northern Ireland Statistics and
Research Agency (NIRSA)

MAP 5.2

This map is part of an All-Island Atlas project developed by AIRO and the ICLRD. The project is part-financed by the European Union's INTERREG IVA programme managed by the Special EU Programmes Body.

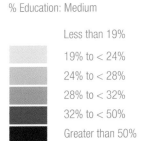

Highest level of education: Medium*
Small Areas (SAs)
*Upper secondary & higher certificate

Dublin City

Belfast City

% Education: Medium

	Less than 19%
	19% to < 24%
	24% to < 28%
	28% to < 32%
	32% to < 50%
	Greater than 50%

	Northern Ireland
	Local Authorities
	Water

	Motorways
	Trunk/Primary Roads
	Secondary Roads
	Streets

Data Source: Central Statistics Office (CSO), Northern Ireland Statistics and Research Agency (NIRSA)

MAP 5.3

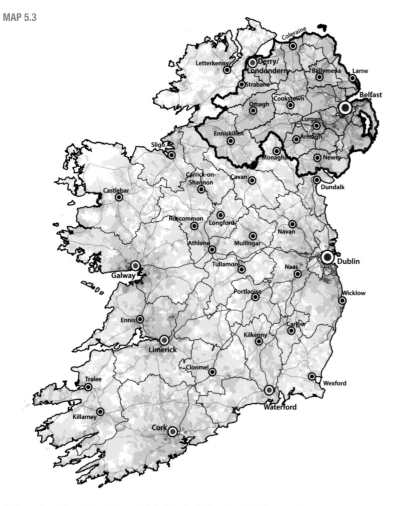

This map is part of an All-Island Atlas project developed by AIRO and the ICLRD. The project is part-financed by the European Union's INTERREG IVA programme managed by the Special EU Programmes Body.

Highest level of education: High*
Small Areas (SAs)
*Third level (degree or higher)

Dublin City

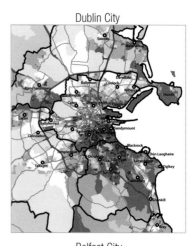

Belfast City

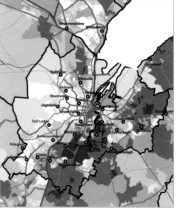

% Education: High

	Less than 12%
	12% to < 20%
	20% to < 29%
	29% to < 40%
	40% to < 55%
	Greater than 55%

	Northern Ireland
	Local Authorities
	Water

	Motorways
	Trunk/Primary Roads
	Secondary Roads
	Streets

Ordnance Survey
Ireland/Government of Ireland
Copyright Permit No. MP 005814

Crown Copyright 2014
Ordnance Survey of Northern Ireland
Permit No. 140029

Data Source: Central Statistics Office
(CSO), Northern Ireland Statistics and
Research Agency (NIRSA)

MAP 5.4

Highest level of education: Apprenticeship
Small Areas (SAs)

Dublin City

Belfast City

This map is part of an All-Island Atlas project developed by AIRO and the ICLRD. The project is part-financed by the European Union's INTERREG IVA programme managed by the Special EU Programmes Body.

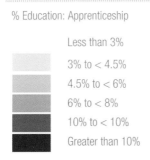

% Education: Apprenticeship

Less than 3%

3% to < 4.5%

4.5% to < 6%

6% to < 8%

10% to < 10%

Greater than 10%

| | Northern Ireland |
| Local Authorities |
| Water |

Motorways

Trunk/Primary Roads

Secondary Roads

Streets

Data Source: Central Statistics Office (CSO), Northern Ireland Statistics and Research Agency (NIRSA)

MAP 5.5

Highest level of education: Vocational
Small Areas (SAs)

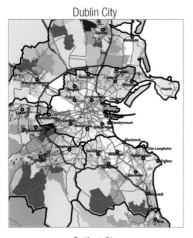

Dublin City

Belfast City

This map is part of an All-Island Atlas project developed by AIRO and the ICLRD. The project is part-financed by the European Union's INTERREG IVA programme managed by the Special EU Programmes Body.

% Education: Vocational

Less than 4%
4% to < 7%
7% to < 9%
9% to < 12%
12% to < 16%
Greater than 16%

Northern Ireland
Local Authorities
Water

Motorways
Trunk/Primary Roads
Secondary Roads
Streets

Ordnance Survey Ireland/Government of Ireland Copyright Permit No. MP 005814

Crown Copyright 2014 Ordnance Survey of Northern Ireland Permit No. 140029

Data Source: Central Statistics Office (CSO), Northern Ireland Statistics and Research Agency (NIRSA)

Chapter 6

Transport

Gavin Daly

6.1 Introduction

The demand for transport and its associated modal split is chiefly a function of population growth, the level of activity in the economy and, importantly, patterns of spatial development. Notwithstanding the recent economic recession, in parallel to the rapid economic and demographic growth across the island of Ireland, transport demand has grown significantly. Taken together, in Ireland commuter numbers for work trips totalled 1.77 million in 2011, an increase of 10.4% when compared with 2001. The corresponding figure for Northern Ireland was 738,659, a growth of 7.5% over the same period. Commuting trips for education totalled just over 1 million in Ireland with 387,687 recorded in Northern Ireland. In total, there are now in excess of 3.9 million people commuting to either work or education on a daily basis.

Obviously, both commuting distances and type of mode of transport varies considerably across the island with distinct spatial patterns evident for both work and education trips. This chapter will highlight some of the main trends for travel to work and education across three categories of transport: private; public and soft modes (walking or cycling). The final section of this chapter will provide a summary of cross-border commuting flows between Ireland and Northern Ireland.

6.2 Private Transport

The increased demand for transport across the island of Ireland has been met predominantly by an overwhelming increase in private modes of travel. In total, 82% of households in Ireland and 77% in Northern Ireland had access to at least one car or van in 2011 and private modes accounted for 73% and 71% of all trips to work in each jurisdiction respectively. In Ireland, 440,000 commuters drove to work in 1991. Twenty years later, this number had risen to 1.26 million persons and only in Dublin City and County (56.5%) did the proportion of

persons using private modes to work fall below 70%, with the highest in Donegal at 80.6%.

Over the period 2001 to 2011, the increase in private commuting was most pronounced in the extended commuting belt of Dublin, such as Laois (+35%), Cavan (+28.3%) and, Meath (+27.5%), and also in Galway (+22.5%). In Northern Ireland a very similar pattern emerged with only Belfast (56.4%) recording commuting by private transport of less than 70% with Magherafelt having the highest overall rate at 78.7%. The highest rates of growth over the period 2001 to 2011 were recorded in Dungannon (+30.3%), Cookstown (+24.7%) and Newry and Mourne (+24.3%). As illustrated in Map 6.1 there is a very clear dependence on private modes of transport to work across the island. In general, rates are lower in the cities and main urban centres with highest rates in rural areas and commuter zones to our main employment destination where rates can be in excess of 82%.

As illustrated in Map 6.2, the trend towards increased commuting by private modes to work was also mirrored in education trips. In Ireland, a continuous rise has been recorded in the use of private modes, with just over half-a-million primary, secondary and third-level students travelling to education by private car. The highest proportion was in more rural counties such as Cork County (62.5%), Galway (61.3%), Tipperary North (61.2%) and Clare (60%), with the lowest proportion in Dublin City and County (36%). However, private travel to education was noticeably lower in Northern Ireland with an average of 39.4% of students using private modes and with only North Down rising above 50%. Ballymena (49.1%), Castlereagh (47.5%) and Carrickfergus (45.2%) also had high percentage shares. Belfast recorded the lowest proportion with just 26.8% of trips to education by private modes.

These trends towards increasing private modes of transport were fundamentally influenced by spatial

patterns of development which during the 2001 to 2011 period saw widespread suburban and diffuse urban sprawl and rapidly expanding commuter belts around larger cities. In Ireland, 75% of the population growth recorded between 1991 and 2011 occurred outside of the larger cities, chiefly in towns of over 1,000 in population but also extensive dispersed settlement in the countryside. Larger cities, however, dominated as employment locations giving rise to extensive long distance work commuting and a consequential impact on education trips. Dublin and its environs, for example, accounted for almost half-a-million jobs in 2011. In Northern Ireland, a similar pattern emerged, intensified by the legacy of the troubles and planning policy which promoted the development of peripheral growth centres and new towns in the wider hinterland of Belfast. These spatial patterns have resulted in embedded car dependency, more demand for road infrastructure (which has tended to further intensify diffuse spatial patterns), and very significant difficulty in providing cost-efficient and effective public transport options given the diverse range and dispersed pattern of journey origins and destinations.

6.3 Soft Modes (Walking/Cycling)

The inevitable result of this growth in car demand has been a substantial decline in the use of other modes. Across the island, the combined share for walking, cycling and public transport to work fell from 20.8% in 2001 to 18.4% in 2011. To put this in perspective, 187,024 more people drove to work over this period but there were 17,133 fewer people who walked or cycled to work and just 1,744 additional public transport commuters. Map 6.3 illustrates the percentage of persons walking or cycling to work in 2011. In Ireland, 11.8% of commuters walked or cycled to work compared with 8.6 % in Northern Ireland, an overall all-island decline of almost 6% between 2001 and 2011. Cycling and walking to work was, as would be expected, heavily concentrated in the larger cities where trip distances are shorter, better infrastructure is available and where there is a lower proportion of car ownership. In 2011, for example, 17.7% of work trips in Dublin City and County were by walking or cycling, an increase of 9.5% over 2001 levels and which reflects the considerable attempts to regenerate inner urban areas and incentivise active modes of transport. The equivalent figure in Belfast was 18.3%, an increase of 12.2%, which also reflects the implementation of urban regeneration policies following the cessation of conflict.

For education, a similar picture emerged with the number of students who walked or cycled to education experiencing a steady decline in the face of rising private modes. In Northern Ireland, 27.8% of students walked or cycled to education which was higher than in Ireland at 21.2%. Again, as illustrated in Map 6.4, walking and cycling to education was heavily concentrated in the larger urban centres, with Belfast, for example, accounting for 30% of all such trips in Northern Ireland and Dublin City and County accounting for 40% of all trips in Ireland.

6.4 Public Transport

The percentage of commuters using public transport (bus or train) also experienced the same steady decline in the face of the increasing use of private modes. In Ireland, the total number of persons using public transport to commute to work increased marginally from 140,381 in 2001 to 144,425 in 2011, an increase of just 2.8%. However, the percentage share declined from 8.7% to 8.1% over the same period. This is despite a very significant government investment in modernising rolling-stock, rail track upgrades and other public transport infrastructure. Map 6.5 again shows that in Ireland the proportion of public transport users was highest in the Dublin area, where more than one in five commuters travelled by bus or train. In Clare, Kerry, Kilkenny, Monaghan, Roscommon and Tipperary just 1.2% of commuters used public transport. In Northern Ireland, however, the number of commuters using public transport actually fell from 47,179 to 44,909, an actual decrease of 4.8% and a share decrease from 6.9% to 6.1%. Belfast had the highest share of public transport users for work trips with 14.5%, while Cookstown and Fermanagh had the lowest shares at just 1.1% and 1.4% respectively.

As discussed above, the numbers of students travelling to education was also dominated by private modes with just 21% of students in Ireland travelling by public transport in 2011. A continuous fall has been recorded

in the use of public transport among primary, secondary and third-level students travelling to education. The lowest proportion of public transport use was in the larger cities, mainly reflecting the above average levels of walking and cycling. The wider extended commuter belts of the main urban centres have a visibly lower use of public transport for school trips, such as in Carlow (13.2%), Kildare (18.9%) and Tipperary North (17.4%). However, as illustrated in Map 6.6, in Northern Ireland the proportion of public transport trips to education was significantly higher with an average of almost 28% using public transport. Interestingly, the use of public transport was more prevalent in more rural western council areas such as Fermanagh (40.7%) and Strabane (36.3%) with lowest proportion in the greater Belfast area. This trend is also replicated on the other side of the border with, for example, Donegal (33.3%), Cavan (31.6%), Leitrim (32.3%) and Monaghan (29.9%) having significantly above average use of public transport for education trips.

6.5 Commuting Flows across the island of Ireland

For the first time, in the 2011 censuses, the place of work or study for persons who travelled across the border from Ireland to Northern Ireland or from Northern Ireland to Ireland was recorded. The outputs from this exercise are available through two main datasets: the Place of Work, School or College Census of Anonymised Records (POWSCAR) database from the CSO; and through the Origin-Destination (OD) Statistics from the Northern Ireland Statistics and Research Agency (NISRA). Unfortunately, at the time of writing the OD data for Northern Ireland was not readily available at a local spatial scale and as such the analysis in this section is based on commuting flows within Ireland and from Ireland to Northern Ireland only. However, a recent joint publication between the CSO and NISRA has provided some headline statistics on commuting flows between both jurisdictions as recorded in Census 2011.

The results from both censuses revealed that a total of 14,800 persons regularly commuted between the jurisdictions for either work or study. Of this total, 44% (6,500) were travelling from Northern Ireland to Ireland and 56% (8,300) were travelling from Ireland to Northern Ireland. The majority of those who commuted to Northern Ireland were resident in the border areas of Donegal, Cavan, Monaghan and Louth. Their places of work or study in Northern Ireland were mainly concentrated in the Belfast and Derry areas with further clusters in larger towns close to the border. The corresponding data for commuters from Northern Ireland also showed that,

while much of the activity is in border areas, the origin of these commuters was more widely spread across Northern Ireland. The destination of these commuters is concentrated in Dublin, with further clusters in the border towns of Letterkenny, Drogheda, Dundalk, Cavan and Monaghan.

An analysis of the POWCAR database from the CSO allows us to undertake some useful analysis on north-bound work based commuting flows. In 2006, 5,277 people (0.28%) of the total workforce residing in Ireland worked in Northern Ireland. By 2011 this had increased to 6,419 people (0.36%). Although this number is relatively low, it is still an important aspect of employment in the Border region, particularly along the Letterkenny-Derry/Londonderry corridor where more than 30% of all local workers in parts of the Inishowen peninsula work in Northern Ireland. Rates are also high in pockets of northwest Cavan, north Monaghan and northeast Louth. Derry City is by far the largest destination for cross-Border workers accounting for 40% of all commuters from Ireland. In comparison, Belfast accounts for only 6.3%. Other cross-Border towns, such as Newry, Enniskillen and Strabane, account for less than 5% each. The remainder of the destinations are scattered around the rest of Northern Ireland.

6.6 Conclusion

The overall transport trends that have emerged over the period 2001 to 2011 pose major questions for future transport and spatial policies and investment planning on the island of Ireland. The growth of geographically dispersed commuter belts around all our major cities, particularly Dublin and Belfast but also Cork, Derry/Londonderry, Galway and Limerick, has been striking. This inherently increases transport demand in the form of longer journey distances and is entirely ill-suited to effective public transport provision, walking or cycling. However, due to spatial patterns of development, even outside of cities, smaller urban areas and rural areas also generate significant levels of transport demand where alternatives to private modes are not a viable option. In Ireland, only about 10% of all kilometres travelled are by modes other than car or van. However, across the island, transport policies have consistently had a stated aim of achieving modal shift away from private modes. For example, the Irish government's 'Smarter Travel' policy introduced in 2009 aims at achieving a 20% reduction on the total share of car commuting to work by 2020 i.e. that half-a-million persons will take alternative means. The Regional Transportation Strategy for Northern Ireland

2002 - 2012 had similar objectives to increase public transport trips, walking and cycling. These objectives have been based on reducing and avoiding the direct costs of congestion and recognising the very high economic cost of increasing road capacity to meet ever growing demand, together with climate change, energy, environmental and public health considerations.

It must now be recognised from the census data that not only have these policies completely failed but trends are actively moving in the opposite direction in both jurisdictions. Even the most recent trends, notwithstanding the economic recession, suggest further growth in the share of private transport modes for commuting and education trips with very limited potential for alternative modes. Dublin and Belfast are increasing

in dominance as employment centres giving rise to further long distance commuting. Dispersed spatial patterns of development have now effectively 'locked-in' acute car dependency and, in the context of the much changed economic circumstances, are very unfavourable to efficient and sustainable transport provision and present serious questions as to whether policies can now meaningfully act to reverse car dependency.

MAP 6.1

Mode of Transport to Work: Private Transport
Small Areas (SAs)

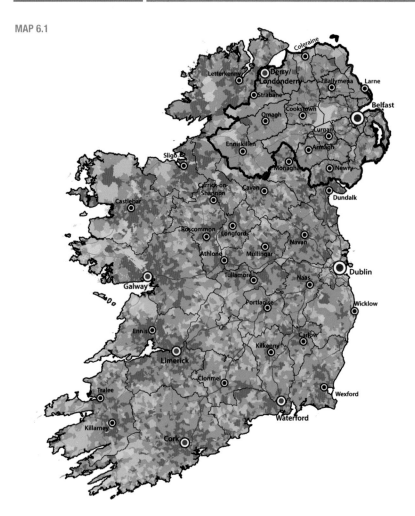

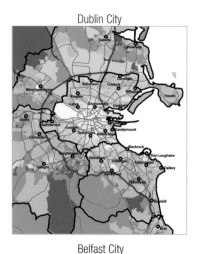

Dublin City

Belfast City

This map is part of an All-Island Atlas project developed by AIRO and the ICLRD. The project is part-financed by the European Union's INTERREG IVA programme managed by the Special EU Programmes Body.

% Mode to Work:
Private

Less than 30%

30% to < 47%

47% to < 60%

60% to < 72%

72% to < 82%

Greater than 82%

Northern Ireland

Local Authorities

Water

Motorways

Trunk/Primary Roads

Secondary Roads

Streets

Ordnance Survey Ireland/Government of Ireland Copyright Permit No. MP 005814

Crown Copyright 2014 Ordnance Survey of Northern Ireland Permit No. 140029

Data Source: Central Statistics Office (CSO), Northern Ireland Statistics and Research Agency (NIRSA)

MAP 6.2

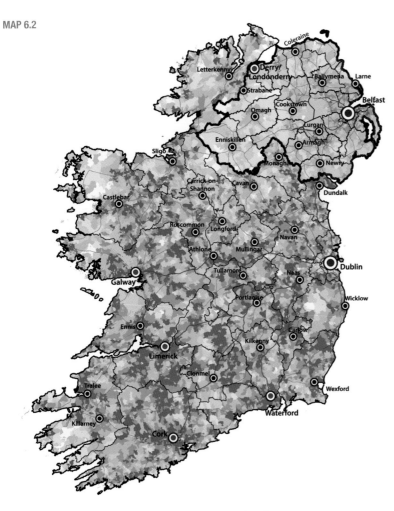

Mode of Transport to Education: Private Transport
Small Areas (SAs)

Dublin City

This map is part of an All-Island Atlas project developed by AIRO and the ICLRD. The project is
part-financed by the European Union's INTERREG IVA programme managed by the Special EU Programmes Body.

Belfast City

% Mode to Education:
Private

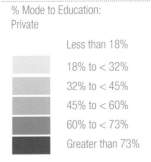

Less than 18%

18% to < 32%

32% to < 45%

45% to < 60%

60% to < 73%

Greater than 73%

Northern Ireland

Local Authorities

Water

Motorways

Trunk/Primary Roads

Secondary Roads

Streets

Ordnance Survey
Ireland/Government of Ireland
Copyright Permit No. MP 005814

Crown Copyright 2014
Ordnance Survey of Northern Ireland
Permit No. 140029

Data Source: Central Statistics Office
(CSO), Northern Ireland Statistics and
Research Agency (NIRSA)

MAP 6.3

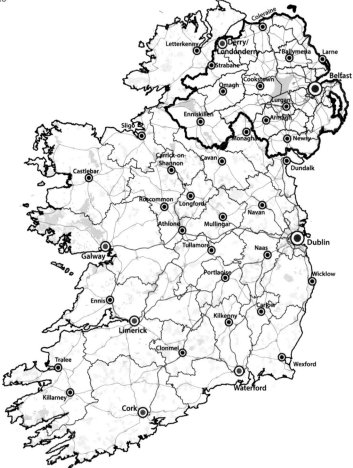

This map is part of an All-Island Atlas project developed by AIRO and the ICLRD. The project is part-financed by the European Union's INTERREG IVA programme managed by the Special EU Programmes Body.

Mode of Transport to Work: Green Mode*
Small Areas (SAs)
*Walking or Bicycle

Dublin City

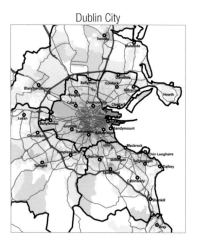

Belfast City

% Mode to Work:
Green Modes

Less than 6%
6% to < 13%
13% to < 22%
22% to < 34%
34% to < 48%
Greater than 48%

Northern Ireland
Local Authorities
Water

Motorways
Trunk/Primary Roads
Secondary Roads
Streets

Ordnance Survey
Ireland/Government of Ireland
Copyright Permit No. MP 005814

Crown Copyright 2014
Ordnance Survey of Northern Ireland
Permit No. 140029

Data Source: Central Statistics Office
(CSO), Northern Ireland Statistics and
Research Agency (NIRSA)

MAP 6.4

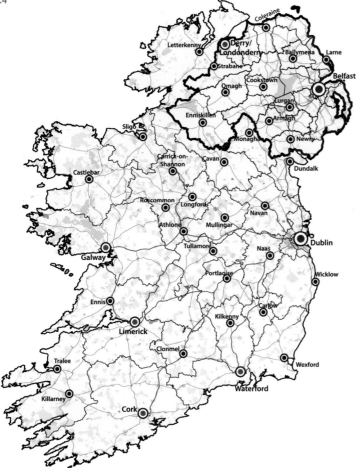

Mode of Transport to Education: Green Mode*
Small Areas (SAs)
*Walking or Bicycle

Dublin City

Belfast City

This map is part of an All-Island Atlas project developed by AIRO and the ICLRD. The project is part-financed by the European Union's INTERREG IVA programme managed by the Special EU Programmes Body.

% Mode to Education:
Green Modes

- Less than 10%
- 10% to < 23%
- 23% to < 36%
- 36% to < 50%
- 50% to < 67%
- Greater than 67%

- Northern Ireland
- Local Authorities
- Water

- Motorways
- Trunk/Primary Roads
- Secondary Roads
- Streets

Data Source: Central Statistics Office (CSO), Northern Ireland Statistics and Research Agency (NIRSA)

MAP 6.5

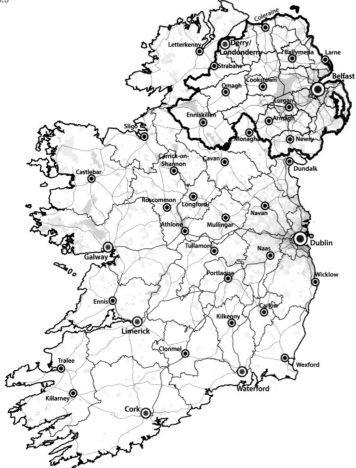

Mode of Transport to Work: Public Transport
Small Areas (SAs)

Dublin City

Belfast City

This map is part of an All-Island Atlas project developed by AIRO and the ICLRD. The project is
part-financed by the European Union's INTERREG IVA programme managed by the Special EU Programmes Body.

% Mode to Work:
Public

Less than 3%

3% to < 8%

8% to < 15%

15% to < 22%

22% to < 31%

Greater than 31%

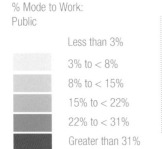

	Northern Ireland
	Local Authorities
	Water

Motorways

Trunk/Primary Roads

Secondary Roads

Streets

Ordnance Survey
Ireland/Government of Ireland
Copyright Permit No. MP 005814

Crown Copyright 2014
Ordnance Survey of Northern Ireland
Permit No. 140029

Data Source: Central Statistics Office
(CSO), Northern Ireland Statistics and
Research Agency (NIRSA)

MAP 6.6

Mode of Transport to Education: Public Transport
Small Areas (SAs)

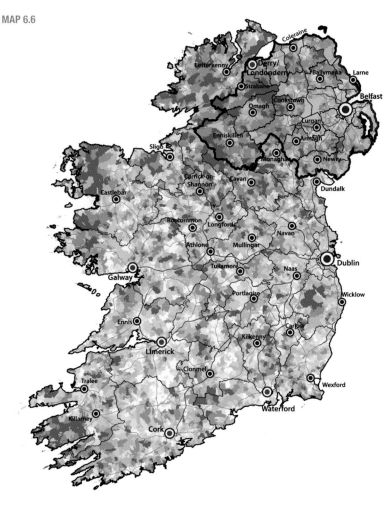

Dublin City

Belfast City

This map is part of an All-Island Atlas project developed by AIRO and the ICLRD. The project is part-financed by the European Union's INTERREG IVA programme managed by the Special EU Programmes Body.

% Mode to Education:
Public

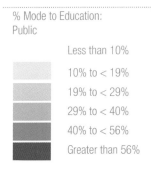

Less than 10%

10% to < 19%

19% to < 29%

29% to < 40%

40% to < 56%

Greater than 56%

Northern Ireland

Local Authorities

Water

Motorways

Trunk/Primary Roads

Secondary Roads

Streets

Data Source: Central Statistics Office (CSO), Northern Ireland Statistics and Research Agency (NIRSA)

MAP 6.7

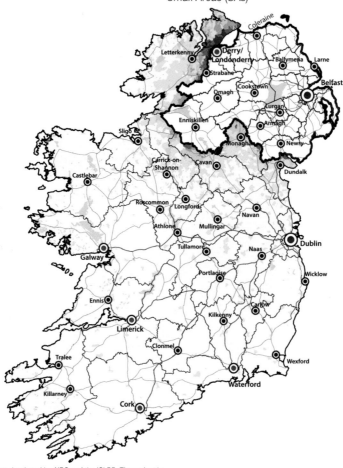

Cross-Border Employment Commuting Flows, 2011
Small Areas (SAs)

This map is part of an All-Island Atlas project developed by AIRO and the ICLRD. The project is
part-financed by the European Union's INTERREG IVA programme managed by the Special EU Programmes Body.

% Cross Border Commuters
Employment

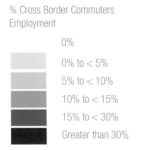

0%

0% to < 5%

5% to < 10%

10% to < 15%

15% to < 30%

Greater than 30%

Northern Ireland

Local Authorities

Water

Motorways

Trunk/Primary Roads

Secondary Roads

Streets

Ordnance Survey
Ireland/Government of Ireland
Copyright Permit No. MP 005814

Crown Copyright 2014
Ordnance Survey of Northern Ireland
Permit No. 140029

Data Source: Central Statistics Office
(CSO), Northern Ireland Statistics and
Research Agency (NIRSA)

Chapter 7

Housing

Prof. Rob Kitchin

7.1 Introduction

The two decades between 1991 and 2011 were a period of enormous change in both Northern Ireland and Ireland. In Northern Ireland, the peace process bought a cessation to The Troubles and ushered in a new period of relative social stability and modest economic growth. In Ireland, the economy diversified and grew rapidly, accompanied by an expansion in population, and a secularisation of society. The most visible manifestation of this growth in economy and population on the Irish landscape was a boom in construction, with large investments into new infrastructure (e.g., transport, utilities), commercial premises (especially offices and retail units), and the building of new houses and apartments.

Even before the economic boom in the South, the geography of housing in Ireland and Northern Ireland differed quite markedly as a function of different political economies, planning systems, and social structure. In 1991 in Northern Ireland, there was a much greater proportion of social housing stock (29.3%) that was administered by a single housing agency, the Northern Ireland Housing Executive, the largest in Europe. Social housing was highly segregated by religion of occupants, as was all housing more generally. The planning system sought to concentrate new housing around existing settlements and to limit the number of new one-off housing. In contrast, in Ireland, the proportion of social housing had been declining from a high of 18.4% in 1961 to 9.7% in 1991, with each local authority responsible for managing and building social housing stock. Moreover, there was a strong cultural emphasis on owner-occupation as the ideal form of tenancy, and a relatively weak, laissez-faire planning system facilitated the construction of rural one-off housing. However, the housing markets in Northern Ireland and Ireland shared the characteristic of being relatively low in cost to buy relative to other European countries.

Between 1991 and the financial crash of 2007, the housing sectors in Ireland and Northern Ireland followed similar trajectories, though the scale of the transformation differed. Both jurisdictions facilitated social housing tenants to buy their own homes. Both jurisdictions experienced very rapid growth in housing prices, with prices in Ireland increasing several-fold in a decade and a half. The average price for a new house in Ireland rose from €66,914 in 1991 to €322,634 in 2007 (a 382% increase), and second-hand homes from €64,122 to €377,850 (a 489% increase), with prices rising even more steeply in Dublin (a 429% increase for new houses and a 551% increase for second-hand ones). Consequently, affordability of housing dropped enormously, as illustrated by the fact that in Q3 1995 the average second-hand house price was 4.1 times the average industrial wage, but by Q2 2007 prices had risen to 11.9 times. There was a growth in private sector renting in both jurisdictions (e.g., in Northern Ireland the number of households privately renting increased from under 30,000 in 1991 to 125,400 in 2011). And whilst the volume of housing stock increased markedly in Northern Ireland, it positively exploded in Ireland, rising by 834,596 units between 1991 and 2011 driven by a rapidly increasing population, household fragmentation, and replacing obsolescent stock. In Ireland, housing completions grew from 21,391 in 1993 to 93,419 in 2006 (Figure 7.1), with Ireland having the highest rate of completions per head of population in Europe in 2007 (over twice the rate of every other country, with the exception of Spain) (Figure 7.2).

Whilst at the time, the property sector, media and politicians celebrated these transformations, arguing that a property bubble had not formed in either jurisdiction and prices and the constitution of stock was adjusting to reflect international norms, this quickly changed when the 2007 global financial crash triggered a radical change in the property sector's fortunes. While both jurisdictions were affected by the general shocks in the global financial system, the collapse in the Irish and

Northern Irish banking sector were exacerbated by their exposure to the home-grown property bubble, inter-bank lending to finance development, and speculation in the Northern market by Southern investors. The result was that in both Northern Ireland and Ireland property prices plunged by circa fifty percent over the next five years (for example, prices fell in the North from an average of £250,000 in Q3 2007 to £129,000 in Q2 2013).

In turn, this exposed and created a series of other issues, including oversupply, extensive mortgage arrears (in Ireland in Q4 2012, 18.2% of all private residential mortgages were in arrears, 11.5% of which were in arrears of more than 90 days; 18.9% of buy-to-let mortgages were in arrears of more than 90 days), widespread negative equity (c. 50% of all residences with a mortgage in 2011), stalled regeneration of social stock, a large social housing waiting list (growing in Ireland 63% between 2007 and 2010 to reach 97,260 households; in Northern Ireland 38,100 households were on the housing list in 2010), and a growing bill to subsidize low income tenants renting privately (96,809 households in June 2011 in Ireland received rent supplement; >60,000 tenants in Northern Ireland received Housing Benefit in 2014). In Ireland, this was complemented by numerous unfinished estates (2,876 in 2011), issues of quality with respect to boom-era stock (e.g., 74 estates, consisting of 12,250 houses, being affected by pyrite; Priory Hall and other developments non-compliant with building regulations).

Seven years after the crash, house prices are now starting to rise again in and around the principal cities, slowly reversing their drastic fall. The number of unfinished estates in Ireland are slowly falling in number as work recommences on them, and the numbers in mortgage arrears are lowering as financial arrangements are agreed between lenders and their bank. However, all of the issues detailed still exist and will take a number of years to fully unwind, especially in rural areas where a large oversupply persists and the local economy is weak.

Figure 7.1: Annual Housing Completions in Ireland, 1993 to 2013

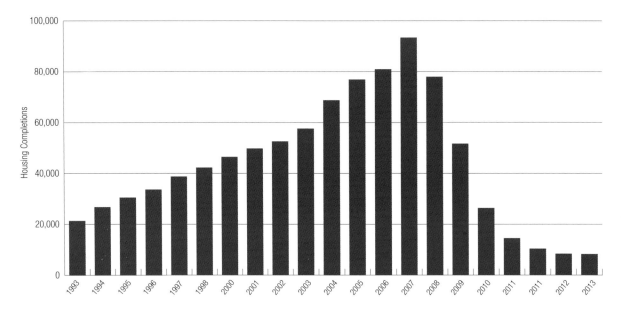

Figure 7.2: Housing Completions per '000 Population, 2007

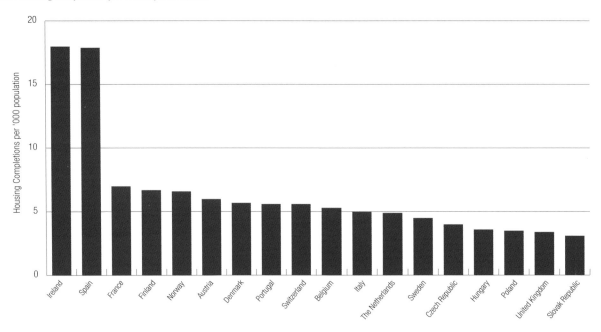

Further, a new crisis is starting to emerge -- somewhat ironically given the cause of the crash -- a shortage of supply in and around Dublin caused by a growing population and very little new construction having taken place over a number of years, leading to a rebound in prices and a two-speed housing market nationally. This last point is illustrative of what the following maps reveal. That the housing sector in Ireland and Northern Ireland varies geographically, with distinct patterns with respect to housing stock, type, and tenure between jurisdictions, between regions, and between urban and rural locales. These patterns are reflective of both long and more recent trends, but means that any analysis of housing on the island of Ireland needs to be sensitive to local context. Moreover, future spatial patterns will develop in different ways dependent on trends in demography and household composition, migration and natural increase, local economy, and planning regulations and local and regional development plans.

7.2 Housing Stock and Vacancy
The total housing stock across the island in 2011 was 2.75 million units, of which 1.99 million are Ireland and 754,000 are in Northern Ireland. In both jurisdictions, stock levels grew significantly from 2001/2 with growth in Northern Ireland of 13.7% (up from 663,572) and in Ireland a massive rise of 36.6% (up from 1,460,053) (Map 7.1). In total, there was a 29.4% increase in stock across the island (626,051). The number of households also rose in the same period (by 12.2% in Northern Ireland and 28.4% in Ireland), but in the case of Ireland households increased by 366,250 whereas stock increased by 534,792 units. The result was a significant mismatch between overall stock (1,994, 845) and households (1,654,208), with the vacancy rate in Ireland increasing from 9.8% in 2001 to 14.5% in 2011 to total 289,581 units. Of these, 59,395 units were classed as holiday homes. Assuming a base vacancy rate of 6% (some stock is always vacant in a normal functioning housing market), this meant that the level of oversupply of housing units vis-a-vis demand in Ireland was approximately 110,000 units. In contrast, in Northern Ireland the vacancy rate increased from 4.6% to 6.2%, slightly above base vacancy.

The increase in stock was geographically uneven in both Ireland and Northern Ireland. In Ireland, the housing stock of seven local authorities increased by over 50% -- Cavan, Fingal, Laois, Leitrim, Longford, Meath, and Wexford. An additional nine local authorities had increases over 40%. In the main these 16 authorities can

be grouped into two main types -- commuting locales around the principal cities, and those local authorities belonging to the Upper Shannon Rural Renewal Scheme, a tax incentive scheme designed to encourage investment in property in the counties of Cavan, Leitrim, Longford, Roscommon and Sligo. The only principal city authority to experience such high growth was Galway City, with a 41.5% increase in stock. Only four local authorities in the South experienced a growth in stock less than 30% - Cork City, Dublin City, Dún Laoghaire-Rathdown, and South Dublin. In contrast, in the North, only four districts experienced a growth in stock above 20% -- Ballymoney, Banbridge, Craigavon, and Limavady, all of which are growing commuter locations for Coleraine, Belfast and Derry. Only three districts have a growth rate less than 10% -- Belfast, Carrickfergus, and Castlereagh, which are older established urban centres, where growth is limited.

Not unsurprisingly, the spatial unevenness in stock growth is matched with an uneven distribution in housing vacancy. Vacancy levels, including holiday homes, was over 20% in nine local authorities in Ireland -- Cavan, Clare, Donegal, Kerry, Leitrim, Longford, Mayo, Roscommon, and Sligo. All nine's vacancy rate excluding holiday homes is over 15% (include map excluding holiday homes). In the case of five of these local authorities, the high levels of stock growth and vacancy is directly related to the Upper Shannon scheme, introduced in 1998. As Figure 7.3 illustrates, house building in those five counties was typically 200-400 houses per year from 1970 to 1996, but grew markedly post-1998, exceeding 1000 units per annum in the mid-2000s. It is little wonder then that these counties experienced the most number of unfinished estates, standardised by the number of households. It is likely that it will take several years for this oversupply to be worked down so that supply and demand are in sync. Only six local authorities in the South had vacancy levels under 10% in 2011, all surrounding Dublin City -- Dún Laoghaire-Rathdown, Fingal, Kildare, Meath, South Dublin, and Wicklow -- where growth in household numbers more closely tracked the building boom. It is these local authorities where house prices have recovered first in Ireland as growing household numbers post-2011 worked down the oversupply, creating a situation where demand started to outstrip supply. In contrast, in Northern Ireland only seven districts had a vacancy level above 8% in 2011 -- Armagh, Coleraine, Dungannon, Fermanagh, Moyle, Newry and Mourne, and Omagh; with five districts with rates below

4.5%, suggesting a shortage of supply in those areas -- Carrickfergus, Castlereagh, Derry, Newtownabbey, and North Down. Due to data comparability issues, a local level housing vacancy map at the All-Island level is unavailable. As an alternative, Map 7.2 details the distribution of unoccupied housing units in 2011 across the island - this is based on housing units with visitors, vacant houses or apartments and holiday homes.

Given current projections concerning household growth in both Ireland and Northern Ireland, and current low levels of construction, one can anticipate that over the remainder of the present decade that the number of units in both jurisdictions will grow quite modestly compared to the previous two decades and that the vacancy level will reduce quite markedly in Ireland. Such a situation is to be welcomed as it starts to harmonise supply and demand. However, the present situation in certain locales in Northern Ireland and Ireland wherein vacancy rates are low suggest the need to create new supply in those areas in order to temper any new bubble in house prices forming.

7.3 Housing Type
In both Ireland and Northern Ireland the majority of households live in houses, with only 9.9% households in Northern Ireland living in apartments/flats (74,409) and 11.3% in the Ireland (183,282). The trend, however, is towards apartment living with the proportion of households residing in such premises rising from 8.5% in Northern Ireland (+18,210) and 8.7% in Ireland (+72,824 households) in 2001/2. This shift towards apartment living is particularly pronounced in and around the principal cities of Belfast, Dublin, Limerick, Galway and Cork, which all have rates of above 15%. In Belfast, the numbers of households living in apartments was 25,991 (20%) an increase of 9,141 from 2001. In Dublin City, apartment living accounted for 34.2% of all households in 2011, increasing by 17,543 households from 2002. The proportion of apartment households was 19.9% for Dún Laoghaire-Rathdown and 16.6% for Fingal, the latter rising from 5% in 2002. In Limerick City it was 15.5%, Galway City 22.5% and in Cork City 17.4%.

In contrast, and perhaps not unsurprisingly, rural areas have much lower proportions of households living in apartments, with rates lower than 4% in a number of districts in Northern Ireland (Ballymoney, Cookstown, Limavady, Magherafelt, and Strabane) and local authorities in Ireland (Cavan, Donegal, Galway County,

Leitrim, Longford, Mayo, Offaly, Roscommon, Tipperary North and South, Waterford County and Wexford). Indeed, only 15 out of 60 districts/local authorities have apartment proportions above 10% (Map 7.3). The housing stock of the island of Island, then, is heavily dominated by houses (Map 7.4).

These houses take different forms - terraced, semi-detached and detached - and varies by size (captured as number of bedrooms in the Census) - (Figure 7.4). The number of households living in semi-detached houses is roughly equivalent, but the proportion of households living in terraced housing is larger in Northern Ireland (25%) than Ireland (17%), with the proportion living in detached houses and bungalows higher in Ireland (42%) than Northern Ireland (38%).

Like apartments, the proportionate mix of detached, semi-detached and terraced housing varies quite significantly across the island, with a marked divide between rural, high commuting areas, and cities. For example, focusing on the South, only six local authorities had detached house rates below 20% (Cork City, Dublin City, Fingal, Limerick City, South Dublin, Waterford

City). In contrast, nine local authorities, all of which are predominately rural, had rates 65% or above (Cavan, Donegal, Galway County, Kerry, Leitrim, Longford, Monaghan, Mayo, Roscommon). Likewise, semi-detached and terraced housing was proportionally higher in cities and commuting counties, with rates above 35% for semi-detached housing in Dún Laoghaire-Rathdown, Fingal, South Dublin, Kildare, Limerick City, Waterford City and Galway City. Similarly, local authorities with over 35% terraced housing include Dublin City, Cork City, Limerick City and Waterford City. In contrast, many rural areas had low rates of terraced housing (<10%). This pattern of housing in the South raises a number of issues concerning, on the one-hand, the servicing and effects of one-off housing with their own septic tanks in rural areas (433,564 households, 26.3% of all households), and issues of density and compactness in urban areas.

7.4 Housing Tenure
The most common form of housing tenure in both Ireland (70%) and Northern Ireland (67%) is owner occupied. Only seven out of sixty districts and local authorities have home ownership rates below 60% -- all of them principal cities (Belfast, Derry, Dublin, Cork, Galway,

Limerick, Waterford) (Map 7.5). Of those homes that are owner occupied, in both Northern Ireland and Ireland, approximately half are owned outright and half are owned with a mortgage.

The geographical pattern of home ownership varies, however. Given the widespread construction and household growth in commuter counties around the principal cities over the past decade, these areas have high proportions of owner-occupiers with a mortgage (and consequently, also high numbers of households in negative equity), with several local authorities/districts with rates above 40% of all households (Map 7.6). In contrast, the proportion of owner occupied without a mortgage exceeds 40% of all households in mostly rural counties, especially in the West (Donegal, Fermanagh, Galway Co, Kerry, Leitrim, Mayo, Moyle, Monaghan, Roscommon, Tipp North, Waterford Co) , where there is a higher proportion of older properties and inter-generation passing on of property (Map 7.7). The micro-scale of Small Areas/Output Areas, reveals that these patterns vary within local authorities/districts. For example Map 7.5 shows the variation in owner occupancy in Dublin and Belfast, with high concentrations in the inner

Figure 7.3: Upper Shannon Housing Completions, 1970 to 2009

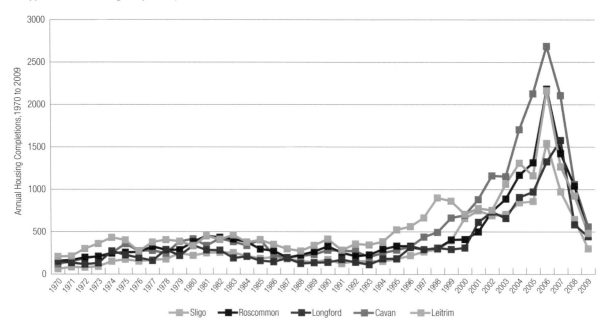

Figure 7.4: Private households by type of accommodation, 2011

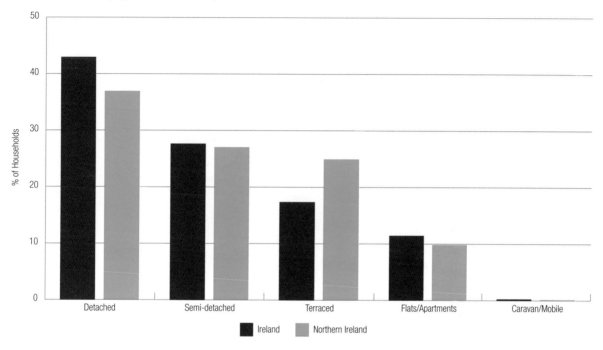

suburbs and low rates in the city centre in both cases. The proportion of owner occupied with a mortgage versus those without mortgages is particularly high in Carrickfergus, Derry, Lisburn, Newtownabbey, Fingal, Kildare, Meath, South Dublin. The numbers of households with a mortgage increased between 2002 and 2011 in all local authorities in Ireland except six (the five city authorities, plus South Dublin), with over a 39% increase in Cavan, Meath, Galway Co, Laois, Leitrim. However, in Northern Ireland, 15 out of the 26 districts experienced an increase, in all cases of less than 13%.

Whilst owner occupied remains by far the most popular form of housing tenancy, between 2001/2 and 2011, there was a reduction in the overall proportion of such households by 7.5% in Ireland and 2% in Northern Ireland. This reduction is predominately due to a growth in private renting in both jurisdictions. In Ireland, the proportion of private renting grew from 12.7% to 20.1% and renting social housing from the local authority grew from 6.9% to 8.7% (a modest increase in a trend of long-term decline). In Northern Ireland, private renting

grew from 9.2% to 17.6%, although the proportion of renters of social housing fell from 21.2% to 14.9% (mostly due to tenants purchasing the property from the Northern Ireland Housing Executive). Even so, social housing still accounts for 46% of all rented properties in Northern Ireland, whereas it is only 30% in Ireland. However, it should be noted that in both Ireland and Northern Ireland, a large proportion of social housing is now provided through the private rental sector supported by rent supplement and the rental accommodation scheme in Ireland and Housing Benefit in Northern Ireland.

Both private rental and social housing vary geographically, both inter- and intra- local authority/ district (Map 7.8 and Map 7.9). In both Ireland and Northern Ireland, they are concentrated into urban settlements. As such, districts/local authorities with more than 20% privately rented households include Belfast, Cork City, Dublin City, Dún Laoghaire-Rathdown, Fingal, Galway City, Limerick City and Waterford City, though there are a few outliers, such as Craigavon, Dungannon,

Omagh. Similarly, the proportion of social housing is above 15% in Belfast, Derry, Lisburn, Strabane, Cork City and Waterford City, with Longford being a rural outlier in the South with a rate of 14.8%. Whilst owner occupation remains the preferred option for many households it seems likely that the trend towards the private rental sector will continue in the short to medium term, especially given the very low volumes of social housing being built since the financial crash.

MAP 7.1

Percentage Change in Housing Stock, 2001/02 to 2011
Local Authorities (Ireland) and Local Government Districts (N. Ireland)

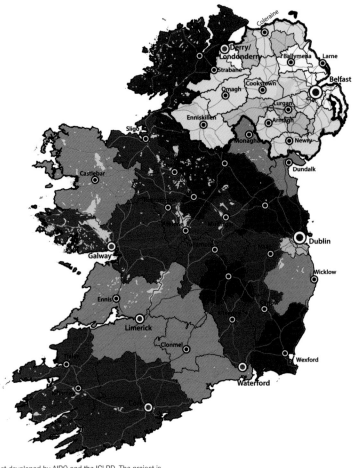

This map is part of an All-Island Atlas project developed by AIRO and the ICLRD. The project is
part-financed by the European Union's INTERREG IVA programme managed by the Special EU Programmes Body.

% Change in Housing Stock
2001/02 to 2011

Less than 13%

13% to < 19%

19% to < 27%

27% to < 37%

37% to < 46%

Greater than 46%

	Northern Ireland		Motorways
	Local Authorities		Trunk/Primary Roads
	Water		Secondary Roads
			Streets

Ordnance Survey
Ireland/Government of Ireland
Copyright Permit No. MP 005814

Crown Copyright 2014
Ordnance Survey of Northern Ireland
Permit No. 140029

Data Source: Central Statistics Office
(CSO), Northern Ireland Statistics and
Research Agency (NIRSA)

MAP 7.2

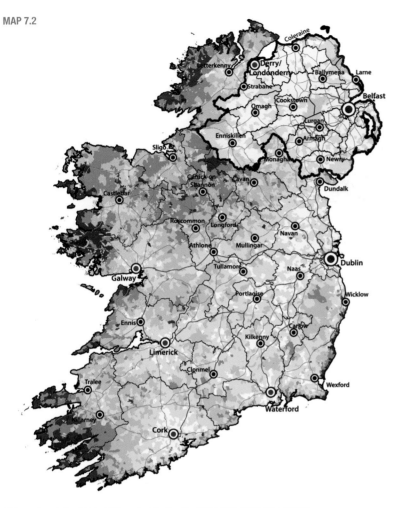

Percentage of Housing Units: Unoccupied*
Small Areas (SAs)
*Visitors, Vacant House/Flat, Holiday Homes

Dublin City

Belfast City

This map is part of an All-Island Atlas project developed by AIRO and the ICLRD. The project is part-financed by the European Union's INTERREG IVA programme managed by the Special EU Programmes Body.

% Housing Units:
Unoccupied

Less than 6%

6% to < 14%

14% to < 24%

24% to < 37%

37% to < 54%

Greater than 54%

Northern Ireland

Local Authorities

Water

Motorways

Trunk/Primary Roads

Secondary Roads

Streets

Ordnance Survey
Ireland/Government of Ireland
Copyright Permit No. MP 005814

Crown Copyright 2014
Ordnance Survey of Northern Ireland
Permit No. 140029

Data Source: Central Statistics Office
(CSO), Northern Ireland Statistics and
Research Agency (NIRSA)

MAP 7.3

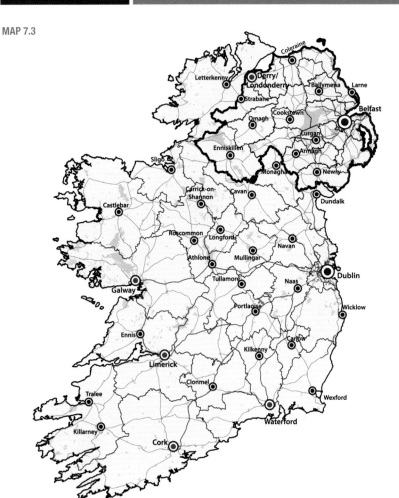

Percentage of Households:
Flats or Apartments
Small Areas (SAs)

Dublin City

Belfast City

This map is part of an All-Island Atlas project developed by AIRO and the ICLRD. The project is
part-financed by the European Union's INTERREG IVA programme managed by the Special EU Programmes Body.

% Households:
Apartments/Flats

Less than 6%
6% to < 20%
20% to < 36%
36% to < 57%
57% to < 81%
Greater than 81%

Northern Ireland
Local Authorities
Water

Motorways
Trunk/Primary Roads
Secondary Roads
Streets

Ordnance Survey
Ireland/Government of Ireland
Copyright Permit No. MP 005814

Crown Copyright 2014
Ordnance Survey of Northern Ireland
Permit No. 140029

Data Source: Central Statistics Office
(CSO), Northern Ireland Statistics and
Research Agency (NIRSA)

MAP 7.4

Percentage of Households: Conventional*
Small Areas (SAs)
*Detached, Semi-D, Bungalow etc

Dublin City

Belfast City

This map is part of an All-Island Atlas project developed by AIRO and the ICLRD. The project is part-financed by the European Union's INTERREG IVA programme managed by the Special EU Programmes Body.

% Households:
House/Terrace/Bungalow

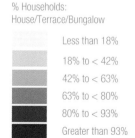

Less than 18%
18% to < 42%
42% to < 63%
63% to < 80%
80% to < 93%
Greater than 93%

Northern Ireland
Local Authorities
Water

Motorways
Trunk/Primary Roads
Secondary Roads
Streets

Ordnance Survey
Ireland/Government of Ireland
Copyright Permit No. MP 005814

Crown Copyright 2014
Ordnance Survey of Northern Ireland
Permit No. 140029

Data Source: Central Statistics Office
(CSO), Northern Ireland Statistics and
Research Agency (NIRSA)

MAP 7.5

This map is part of an All-Island Atlas project developed by AIRO and the ICLRD. The project is part-financed by the European Union's INTERREG IVA programme managed by the Special EU Programmes Body.

Percentage of Households: Owner Occupied
Small Areas (SAs)

Dublin City

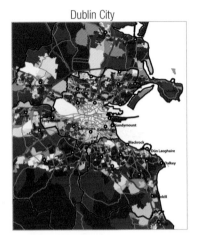

Belfast City

% Households:
Owner Occupied

Less than 20%
20% to < 40%
40% to < 60%
60% to < 70%
70% to < 85%
Greater than 85%

Northern Ireland
Local Authorities
Water

Motorways
Trunk/Primary Roads
Secondary Roads
Streets

MAP 7.6

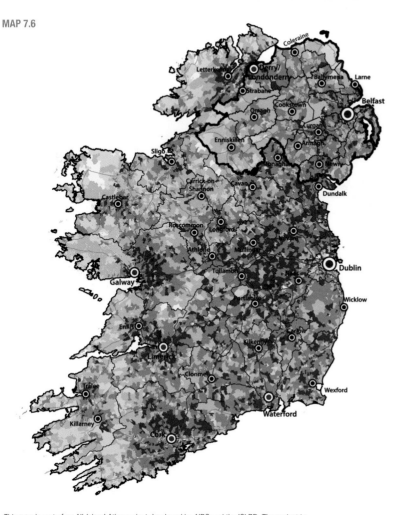

This map is part of an All-Island Atlas project developed by AIRO and the ICLRD. The project is part-financed by the European Union's INTERREG IVA programme managed by the Special EU Programmes Body.

Percentage of Households:
Owner Occupied (Mortgage)
Small Areas (SAs)

Dublin City

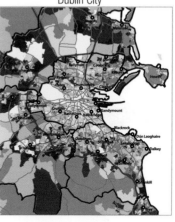

Belfast City

% Households:
Owner Occupied - Mortgage

- Less than 15%
- 15% to < 25%
- 25% to < 35%
- 35% to < 45%
- 45% to < 60%
- Greater than 60%

Northern Ireland
Local Authorities
Water

Motorways
Trunk/Primary Roads
Secondary Roads
Streets

Ordnance Survey
Ireland/Government of Ireland
Copyright Permit No. MP 005814

Crown Copyright 2014
Ordnance Survey of Northern Ireland
Permit No. 140029

Data Source: Central Statistics Office
(CSO), Northern Ireland Statistics and
Research Agency (NIRSA)

MAP 7.7

This map is part of an All-Island Atlas project developed by AIRO and the ICLRD. The project is part-financed by the European Union's INTERREG IVA programme managed by the Special EU Programmes Body.

Percentage of Households:
Owner Occupied (No Mortgage)
Small Areas (SAs)

Dublin City

Belfast City

% Households:
Owner Occupied - No Mortgage

- Less than 10%
- 10% to < 25%
- 25% to < 35%
- 35% to < 45%
- 45% to < 60%
- Greater than 60%

Northern Ireland
Local Authorities
Water

Motorways
Trunk/Primary Roads
Secondary Roads
Streets

Ordnance Survey
Ireland/Government of Ireland
Copyright Permit No. MP 005814

Crown Copyright 2014
Ordnance Survey of Northern Ireland
Permit No. 140029

Data Source: Central Statistics Office
(CSO), Northern Ireland Statistics and
Research Agency (NIRSA)

MAP 7.8

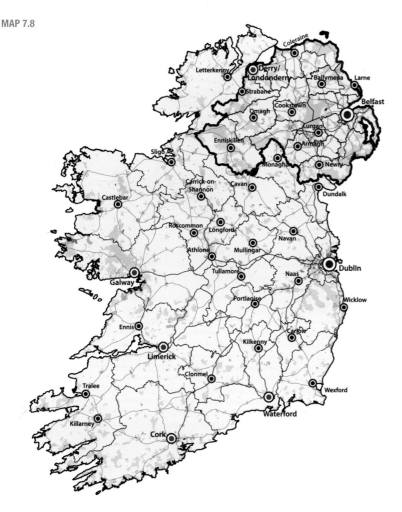

This map is part of an All-Island Atlas project developed by AIRO and the ICLRD. The project is
part-financed by the European Union's INTERREG IVA programme managed by the Special EU Programmes Body.

Percentage of Households:
Private Rented
Small Areas (SAs)

Dublin City

Belfast City

% Households:
Private Rented

Less than 10%

10% to < 20%

20% to < 30%

30% to < 50%

50% to < 70%

Greater than 70%

Northern Ireland

Local Authorities

Water

Motorways

Trunk/Primary Roads

Secondary Roads

Streets

Ordnance Survey
Ireland/Government of Ireland
Copyright Permit No. MP 005814

Crown Copyright 2014
Ordnance Survey of Northern Ireland
Permit No. 140029

Data Source: Central Statistics Office
(CSO), Northern Ireland Statistics and
Research Agency (NIRSA)

MAP 7.9

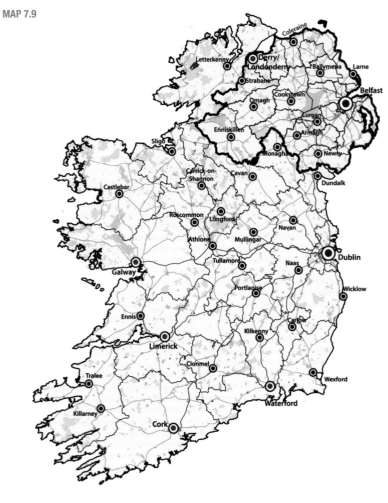

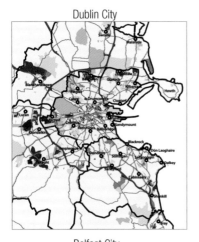

Percentage of Households:
Social Housing
Small Areas (SAs)

Dublin City

Belfast City

This map is part of an All-Island Atlas project developed by AIRO and the ICLRD. The project is
part-financed by the European Union's INTERREG IVA programme managed by the Special EU Programmes Body.

% Households:
Social Housing

Less than 6%

6% to < 16%

16% to < 30%

30% to < 45%

45% to < 70%

Greater than 70%

	Northern Ireland
	Local Authorities
	Water

Motorways

Trunk/Primary Roads

Secondary Roads

Streets

Ordnance Survey
Ireland/Government of Ireland
Copyright Permit No. MP 005814

Crown Copyright 2014
Ordnance Survey of Northern Ireland
Permit No. 140029

Data Source: Central Statistics Office
(CSO), Northern Ireland Statistics and
Research Agency (NIRSA)

Chapter 8

Nationality and Ethnicity

Prof. Mark Boyle

8.1 Introduction

Migration is a term that denotes the movement of people from one location to another. Internal migration is the movement of people from one city or region within a country to another city or region also in that country. International migration is the movement of people from one country to another. Whilst the island of Ireland has a long history of emigration (creating one of the world's largest diasporas estimated at 70 million; there are currently 1.2 million Irish born and 3.1 million Republic of Ireland passport holders living overseas), less well known is its history as a destination for immigrants. And yet during the period of the Celtic Tiger (1993 to 2007) Ireland became a net importer of people, whilst the Peace Process (especially following the Good Friday Agreement (GFA) of 1998) resulted in more migrants from a wider range of backgrounds moving to Northern Ireland. A significant minority of those who moved to Ireland in this period were Irish and British nationals respectively (and their siblings) who were returning "home" after dwelling overseas for a number of years, in some cases decades. But many foreign nationals also came to the island, not least from the new EU accession countries. Moreover, in the early 2000s both Ireland and the United Kingdom had historically unprecedented numbers of applications for asylum. The result: today, the island of Ireland has a rich immigrant community and is home to migrants who hail from over 200 countries around the world. But of course the global economic crises from 2007 has further complicated this picture. Emigration of Irish and British nationals from both Ireland and Northern Ireland has once more returned as a significant demographic phenomenon. Moreover, a proportion of recent immigrants, especially from the EU accession countries, have returned to their source countries or moved from both Ireland and Northern Ireland to an alternative third country. Growth in migrant numbers has levelled off and in the case of the Ireland net emigration has returned. At the same time, the number of applications for asylum has dropped dramatically in both Ireland and the United Kingdom.

Against this backdrop, this chapter presents a profile of the migrant stock in both Ireland and Northern Ireland at Census Day 2011.

8.2 Migration, Ethnicity and Place of Birth

Advances in communication means that many migrants now move in a transient, circular, and more nomadic way than hitherto, and lead transnational existences that entail moving into and out of homelands for short periods. It is often difficult, then, to distinguish between long term migrants and short term migrants, and even between migrants, business travellers and long-stay tourists. Census 2011 included all those who consider themselves to be 'usually resident' in either Ireland or Northern Ireland, in other words who regard their address in Ireland or Northern Ireland to be their principal address. Meanwhile migration can be voluntary or forced. Voluntary migration occurs when people, of their own free will and perhaps in search of a better life, migrate to a new destination which, at least for a while, becomes their new home. Forced migration, in contrast, occurs when people are forced to flee from their homelands due to a humanitarian disaster caused by human actions (war, persecution, famine, etc.) or a natural hazard event (drought, hurricane, earthquake, tsunami, etc.). Forced migration can generate applications for asylum in host societies (permanent full citizenship) from those displaced. If successful, such "asylum seekers" are then reclassified as "refugees". Asylum applicants were included in Census 2011 as part of the "official" census populations, but their status as applicants was not recorded in a way which enables a detailed statistical profile of their socio-economic and geographic character to be drawn.

Migrants are often classified according to their place of birth or country of origin. For the present purposes it is useful to distinguish between migrants whose place of birth is in the "Ireland", "Northern Ireland", "Rest of UK", "Rest of EU27", and "Rest of World". Place of birth can be a proxy for 'nationality', 'ethnicity' and 'race' but it is important not to elide these terms. Nationality is a socially-defined category of people who identify with each other based on a common ancestry. Ethnicity meanwhile is a broader category which encompasses people who share a common culture, language, religion, and/or set of beliefs. Race finally refers to a person's DNA/physical appearance, and is recognized colloquially in terms of skin color, eye color, hair color, bone/jaw structure, etc. Whilst speaking in terms of 'ethnicity', Census 2011 in-fact conflate these three categories. In Ireland people are classified as one of: "White Irish", "White Irish Traveller", "Any other White background", "Black African", "Any other Black background", "Asian Chinese", "Any other Asian background", "Mixed ethnicity", and "Any other not specified above including mixed ethnicity". In Northern Ireland meanwhile the following categories are applied: "White", "Irish Traveller", "Black Caribbean", "Black African", "Black", "Chinese", "Indian", "Pakistani", "Bangladeshi", "Other Asian", "Mixed ethnic group", and "Any other not specified above". For the present purposes a common set of categories between both jurisdictions will be applied: "White", "Irish Traveller", "Black", "Asian", and "Other/Mixed".

8.3 Migration to Ireland in an International Context

According to the United Nations Population Division, in 2013, 232 million people, or 3.2% of the world's population, were international migrants, up from 175 million in 2000 and 154 million in 1990. According to the 2011 Census, in Northern Ireland 11% of usual residents or 202,112 people were born outside the jurisdiction, while in Ireland the equivalent figure was 766,780 people or 17%. Northern Ireland and especially the Ireland then, house a significantly larger share of international migrants than might be expected were international averages to apply.

Globally, migration between countries in the Global South (an imagined region of the world comprising the more underdeveloped countries, most of which reside in the Southern Hemisphere, hence the reference to "South") is as important as migration between countries in the Global South and countries in the Global North (an imagined region of the world comprising the more developed countries, most of which reside in the Northern Hemisphere, hence the reference to "North"). But since 1990, South–North migration and to a lesser extent North to North migration have been the main drivers of global migration trends and it is likely in future that these migration corridors will come to dominate. In 2013, of the 135.6 million migrants who resided in the Global North, 60% came from countries of origin in the Global South and 40% countries in the Global North. Asia is the largest source region for international migrants. In 2013, more Asians lived overseas than migrants from any other continent. Europe remains the greatest magnet of all the world's continents housing 72 million international migrants in 2013. Meanwhile, half of all international migrants were living in 10 countries, with the United States hosting the largest number (45.8

million), followed by the Russian Federation (11 million); Germany (9.8 million); Saudi Arabia (9.1 million); United Arab Emirates (7.8 million); United Kingdom (7.8 million); France (7.4 million); Canada (7.3 million); Australia (6.5 million); and Spain (6.5 million).

Migration to the island of Ireland forms part of this wider picture. In both Ireland and Northern Ireland however, a significantly disproportionate preponderance of migrants arrive from countries located in the Global North and in particular from EU 27 countries (Figure 8.1). In contradistinction to erstwhile popular conceptions that migrants from the third world threaten to "flood" the island, Ireland in-fact houses proportionately significantly fewer migrants from the Global South than is true for the countries of the Global North more generally. In the case of Ireland, of the resident population born elsewhere 542,780 (71% of the foreign born population or 12.5% of the total population of Ireland) was born in another EU27 country whilst 224,000 (29% of the foreign born population or 4.7% of the total population of Ireland) were extra EU27 residents. Meanwhile In the case of Northern Ireland, of the resident population born

elsewhere 165,912 (82% of the foreign born population or 12.5% of the total population of Northern Ireland) were from another EU27 country whilst 36,200 (18% of the foreign born population or 2% of the total population of Northern Ireland) were born in a non EU 27 country.

Firstly, of course there exist migrations of a sort between both Ireland and Northern Ireland. Of the total of 202,112 people usually resident in Northern Ireland in 2011 who were born elsewhere 37,900 were born in the Ireland (representing over 20% of the 'foreign' born stock and 2.1% of the total population of Northern Ireland) (Maps 8.1 and 8.2). Meanwhile of the 766,780 people usually resident in Ireland in 2011 who were born elsewhere, 58,500 were born in Northern Ireland (representing 8% of the foreign born stock and 1.3% of the total population of Ireland). In both cases migrants have tended to relocate to immediately adjacent border counties. Meanwhile included in the foreign born stock resident in the Ireland are 241,985 'returning' Irish nationals born abroad (178,945 born in the UK, 16,703 in US, 3,220 in Australia, 2,524 in Canada and 2,440 in South Africa). Principal source countries for foreign

Figure 8.1: Place of Birth for Ireland and Northern Ireland residents, 2011

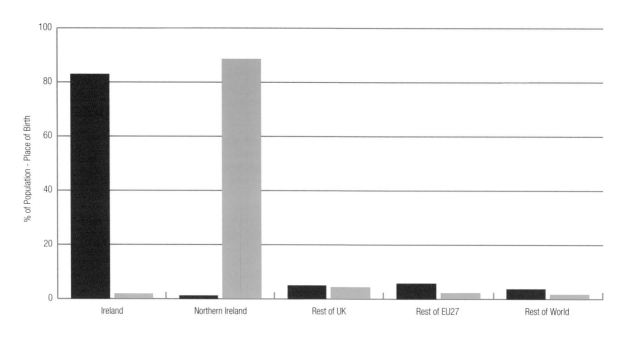

nationals now living in Ireland include: the UK (212,286), Poland (115,193), Lithuania (34,897), United States (27,726), Latvia (19,989), Nigeria (19,780), Romania (17,995), Scotland (17,871) and India 17,856. In the case of Northern Ireland important countries of origin include England and Wales (67,269), Poland (19,658), Scotland (15,455), Lithuania (7,341), India (4,796), United States (4,251), Germany (3,908), the Philippines (2,947) and Slovakia (2,681).

In 2013, the total number of refugees in the world was estimated at 15.7 million, about 7% of all international migrants (United Nations Population Division, 2013). In 2012 in NI there were 240 applications for asylum of which 80 application (or 33 per cent) were successful. Owing to the fact that an unknown number of refugees relocate from Northern Ireland to the UK mainland there exist no data on the total refugee population in Northern Ireland. In Ireland in 2013 there were 946 applications for asylum, 128 (or 13 per cent) of which were granted refugee status (with 13 outstanding). At 9,629 the Irelands total refugee population comprises only 0.01% of the country's population. Ireland (and most likely to a greater extent) Northern Ireland attract less applications for asylum than one might expect given their population size and house fewer refugees than one might expect given the scale of their immigration streams. Not surprisingly asylum applicants are drawn from a wider geographical base than the rest of the foreign born population, the most frequent countries of origin being China, Nigeria, Pakistan and the Democratic Republic of Congo, Afghanistan, Iraq, Syria and Somalia.

Given the origins of their migrant populations, it is unsurprising that both Northern Ireland and Ireland remain ethnically homogenous, with the ethnic category "White" dominating (NI: 98.2% White, with Ireland: 93.6% White), followed by "Black" (Northern Ireland: 0.2% (3,616) with Ireland: 1.4% (65,078)), "Asian" (Northern Ireland: 1.1% (19,130) with Ireland: 1.9% (84,690)), Other Mixed (Northern Ireland: 0.5% (6,505) with Ireland 0.9% (41,220)) and finally Irish Traveller (Northern Ireland: 0.1% (1,301) with Ireland: 0.7% (29,495)). Moreover, not all non White ethnic peoples can be regarded as 'foreign'. Virtually all Irish Travellers living in Ireland and 98% of travellers dwelling in Northern Ireland were born in Ireland or the UK. While the majority of non White ethnic groups were born outside of the EU 27, 34% of those in the Black ethnic groups in Ireland were born in the country, whilst 17% of those citing Asian ethnicity were born in Ireland. This contrasts with

Northern Ireland where 21% of residents with an Asian background were born in Northern Ireland compared to 15% Black ethnic residents. Those in the 'other ethnic group' category including people of mixed ethnicity were more likely than any other non-white ethnic group to be born in either Ireland or Northern Ireland.

8.4 Geographical Spread of Migrants and Ethnic Minorities on the Island

Many migrant groups, out of both choice and necessity, cluster into ethnic neighbourhoods, enclaves, and ghettoes. These clusters, agglomerations and concentrations provide protection and security for new migrants. They can serve as a beachhead into new communities and an elevator of immigrant advancement. They can help new migrants find their feet, secure cultural comforts, leverage practical help, and plant roots more quickly. But sometimes they reflect the limited choices open to migrants in the housing and labour market and are artefacts of exclusion. Sometimes ethic clusters also permit some migrants to live in a goldfish bowl, looking out at the wider community but having no contact with that community. In the national or ethnic cluster there can exist people who can survive without learning the indigenous language or assimilating and integrating into mainstream society. There exists a debate, then, as to the degree to which ethnic neighbourhoods serve to accelerate integration or retard and even render it impossible.

Whilst it is debateable whether the notion of the ethnic neighbourhoods, enclaves, clusters and ghettoes can be applied to migrant and ethnic groups residing on the island of Ireland it is the case that most groupings show evidence of clustering and concentration. In Ireland, Galway City has the highest proportion of residents born elsewhere (25% of all residents), followed by Fingal, Donegal, Dublin City and Monaghan (all> 20% of the total population). Meanwhile some counties (especially Kilkenny, Offaly, Limerick, Waterford and North Tipperary) house comparatively low levels of migrant populations. Meanwhile, in Northern Ireland the highest rates of foreign born populations are to be found in Fermanagh and Dungannon (>16% foreign born) whilst Ballymoney and Magherafelt displayed the lowest rates (around 7%). With respect to residents born in an the Rest of UK, interestingly peripheral, holiday and rural locations are preferred in counties such as Mayo, Leitrim, Donegal, North Down and Roscommon. In contrast populations residing outside of Ireland/UK tend to record their highest rates in the main cities and urban areas

(Galway City, Fingal and Dublin City (>16% foreign born), highest in Northern Ireland is Dungannon 10.3%). But interestingly Belfast with only 6.5% of its population born outside the UK and Northern Ireland/Ireland is the exception. Meanwhile Polish migrants, concentrate in urban centres across the island and display distinctive spatial patterns inside the main cities. Moreover those with Black ethnicity show evidence of concentrating in Fingal, Galway City, South Dublin and Louth in Ireland and Dungannon in Northern Ireland. Moreover the Black population displays a very clear spatial distribution in Dublin City, dwelling in a number of specific peripheral locations. Asian migrants also chose to live in the principal urban locations (Dublin, Galway, Belfast) and for the most part cluster into central locations in these cities. Travellers meanwhile concentrate in only 17% of Small Areas across the island with high rates in Tuam, Rathkeale, Athlone, Dublin periphery, Galway County, Longford, and Galway City. See Maps 8.3 to 8.9 for geographical distributions across the Small Areas in Ireland.

8.5 The impact of Migration on Northern Ireland and Ireland

The impact of migration on both Northern Ireland and Ireland has generated significant topical discussion. But alas a balanced debate on the merits and demerits of immigration has proven difficult. Appropriate discussions on the need for managed migration have at times given way to a post crash austerity fuelled politics which has at its worst moments pandered to racist and xenophobic attitudes and more recently to forms of Islamophobia. And so it is no surprise that many countries are trying to stop migration at its source, tightening their immigration and asylum rules, and toughening their border controls. Increasingly, both Ireland and Northern Ireland are being confronted by debates of this ilk. It is necessary to cut through at times misinformed public opinion and to establish a debate on managed migration which is grounded in fact. Clearly the impact of migrants varies depending upon the country of origin of migrants, their chosen destination, their age and skill levels, and how capable host societies are at harnessing these skills. Migrants can bring skills which are absent or in short supply in the labour market. They are often prepared to do jobs that locals are less keen to do. In regions where population ageing is an issue migrants can correct imbalanced population profiles and dependency ratios and help to pay for health care and pensions. Many work as carers tending the children of workers in host countries and thereby freeing both women and men to join the labor force. Immigrants often bring with them energy and innovation and are disproportionately more likely to become entrepreneurs and to start up their own business. They also enrich the culture of host societies and promote cultural diversity and cosmopolitanism. On the other hand, because of their flexibility migrants can depress the wages paid to other (normally unskilled) workers. Migration can lead to greater unemployment, can create ethic and racial tensions, and can bring security threats. They can also place pressure on scarce health, education, and housing resources.

Either way, it is clear that migrants can contribute most to host societies when then become integrated into the fabric of those societies. Integration with host societies can take one of four principal forms. Social integration occurs when migrants forge new social networks in the host society and secure equal access to housing, educational, health, and recreation facilities. Economic integration occurs when migrants plant new roots in the host society and secure employment. Through upward socioeconomic mobility they begin to achieve parity with the socioeconomic profile of the host population. Political integration occurs when migrants gain full citizenship rights in destination countries and participate as equal members in the political life of the nation (vote in elections, stand for elections, participate in public debate, and so on). Finally, cultural integration occurs when migrants are permitted to celebrate their own cultures, traditions, and customs whilst coming to appreciate the cultural practices of the local population and local ways of life (clothing, dance, religion, food, national belonging, etc.).

Whether Ireland and Northern Ireland will develop into more or less hospitable places for immigrants to live and work remains to be seen.

MAP 8.1

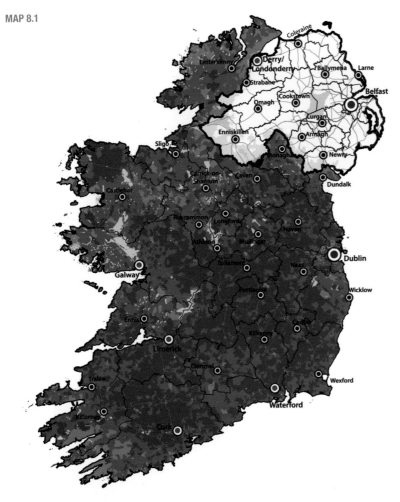

This map is part of an All-Island Atlas project developed by AIRO and the ICLRD. The project is part-financed by the European Union's INTERREG IVA programme managed by the Special EU Programmes Body.

Place of Birth - Republic of Ireland
Small Areas (SAs)

Dublin City

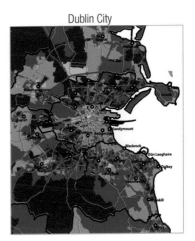

Belfast City

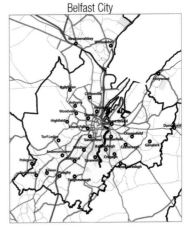

Place of Birth
% Republic of Ireland

Less than 5%

5% to <20%

20% to <50%

50% to <80%

80% to <90%

Greater than 90%

Northern Ireland

Local Authorities

Water

Motorways

Trunk/Primary Roads

Secondary Roads

Streets

MAP 8.2

Place of Birth - Northern Ireland
Small Areas (SAs)

Dublin City

Belfast City

This map is part of an All-Island Atlas project developed by AIRO and the ICLRD. The project is part-financed by the European Union's INTERREG IVA programme managed by the Special EU Programmes Body.

Place of Birth
% Northern Ireland

Less than 5%
5% to <20%
20% to <50%
50% to <80%
80% to <90%
Greater than 90%

Northern Ireland
Local Authorities
Water

Motorways
Trunk/Primary Roads
Secondary Roads
Streets

Ordnance Survey
Ireland/Government of Ireland
Copyright Permit No. MP 005814

Crown Copyright 2014
Ordnance Survey of Northern Ireland
Permit No. 140029

Data Source: Central Statistics Office
(CSO), Northern Ireland Statistics and
Research Agency (NIRSA)

MAP 8.3

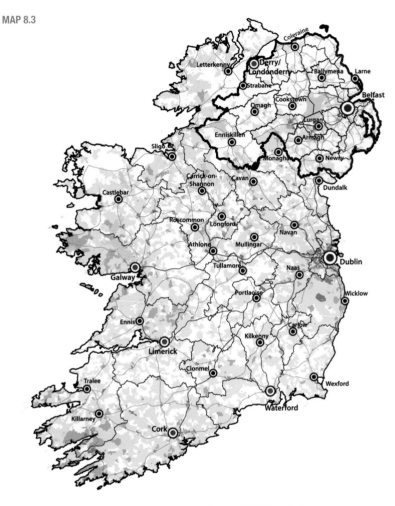

This map is part of an All-Island Atlas project developed by AIRO and the ICLRD. The project is
part-financed by the European Union's INTERREG IVA programme managed by the Special EU Programmes Body.

Place of Birth - Non RoI/UK*
Small Areas (SAs)
*Outside RoI, NI and rest of UK

Dublin City

Belfast City

Place of Birth
% Outside Ireland and UK

Less than 2%

2% to <7%

7% to <15%

15% to <30%

30% to <50%

Greater than 50%

Northern Ireland

Local Authorities

Water

Motorways

Trunk/Primary Roads

Secondary Roads

Streets

Ordnance Survey
Ireland/Government of Ireland
Copyright Permit No. MP 005814

Crown Copyright 2014
Ordnance Survey of Northern Ireland
Permit No. 140029

Data Source: Central Statistics Office
(CSO), Northern Ireland Statistics and
Research Agency (NIRSA)

MAP 8.4

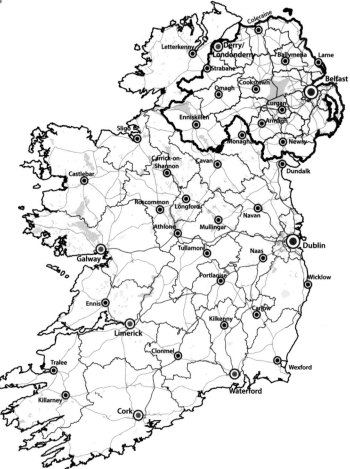

Ethnicity - Black or Black Irish, 2011
Small Areas (SAs)

Dublin City

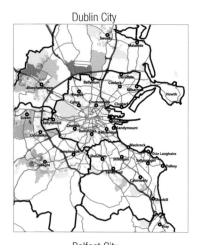

Belfast City

This map is part of an All-Island Atlas project developed by AIRO and the ICLRD. The project is
part-financed by the European Union's INTERREG IVA programme managed by the Special EU Programmes Body.

Ethnicity
% Black or Black Irish

Less than 1.5%

1.5% to <4%

4% to <9%

9% to <17%

17% to <30%

Greater than 30%

Northern Ireland

Local Authorities

Water

Motorways

Trunk/Primary Roads

Secondary Roads

Streets

Ordnance Survey
Ireland/Government of Ireland
Copyright Permit No. MP 005814

Crown Copyright 2014
Ordnance Survey of Northern Ireland
Permit No. 140029

Data Source: Central Statistics Office
(CSO), Northern Ireland Statistics and
Research Agency (NIRSA)

MAP 8.5

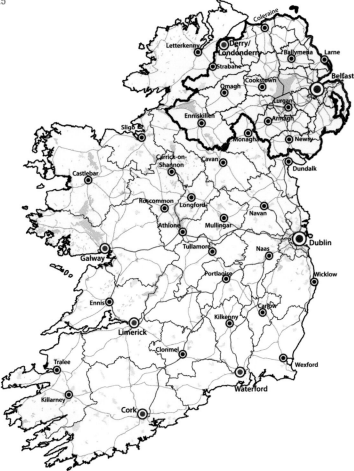

Ethnicity - Asian or Asian Irish
Small Areas (SAs)

Dublin City

Belfast City

This map is part of an All-Island Atlas project developed by AIRO and the ICLRD. The project is part-financed by the European Union's INTERREG IVA programme managed by the Special EU Programmes Body.

Ethnicity
% Asian or Asian Irish

Less than 1.5%
1.5% to <5%
5% to <10%
10% to <17%
17% to <30%
Greater than 30%

Northern Ireland
Local Authorities
Water

Motorways
Trunk/Primary Roads
Secondary Roads
Streets

Ordnance Survey
Ireland/Government of Ireland
Copyright Permit No. MP 005814

Crown Copyright 2014
Ordnance Survey of Northern Ireland
Permit No. 140029

Data Source: Central Statistics Office
(CSO), Northern Ireland Statistics and
Research Agency (NIRSA)

MAP 8.6

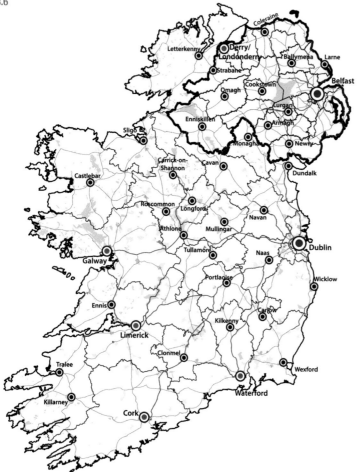

Ethnicity - Irish Traveller, 2011
Small Areas (SAs)

Dublin City

Belfast City

This map is part of an All-Island Atlas project developed by AIRO and the ICLRD. The project is
part-financed by the European Union's INTERREG IVA programme managed by the Special EU Programmes Body.

Ethnicity
% Travellers

Less than 1.5%

1.5% to <5%

5% to <10%

10% to <20%

20% to <40%

Greater than 40%

Northern Ireland

Local Authorities

Water

Motorways

Trunk/Primary Roads

Secondary Roads

Streets

Ordnance Survey
Ireland/Government of Ireland
Copyright Permit No. MP 005814

Crown Copyright 2014
Ordnance Survey of Northern Ireland
Permit No. 140029

Data Source: Central Statistics Office
(CSO), Northern Ireland Statistics and
Research Agency (NIRSA)

MAP 8.7

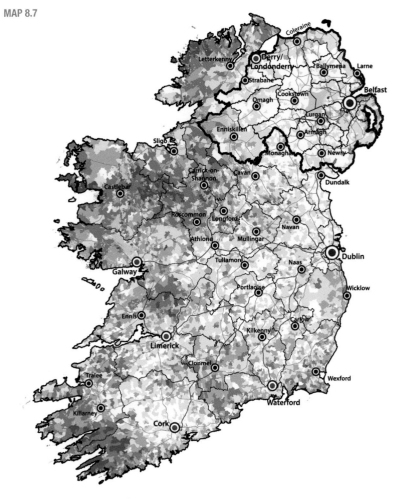

Place of Birth - Rest of UK*
Small Areas (SAs)
*England, Wales and Scotland

Dublin City

Belfast City

This map is part of an All-Island Atlas project developed by AIRO and the ICLRD. The project is
part-financed by the European Union's INTERREG IVA programme managed by the Special EU Programmes Body.

Place of Birth
% UK (Exc NI)

Less than 3.5%

3.5% to <6%

6% to <9%

9% to <15%

15% to <35%

Greater than 35%

Northern Ireland	
Local Authorities	
Water	

Motorways

Trunk/Primary Roads

Secondary Roads

Streets

Ordnance Survey
Ireland/Government of Ireland
Copyright Permit No. MP 005814

Crown Copyright 2014
Ordnance Survey of Northern Ireland
Permit No. 140029

Data Source: Central Statistics Office
(CSO), Northern Ireland Statistics and
Research Agency (NIRSA)

MAP 8.8

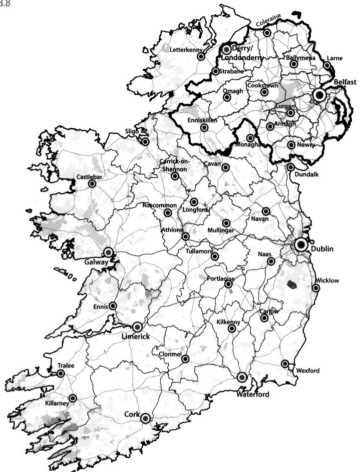

Place of Birth - Rest of EU27
Small Areas (SAs)

Dublin City

Belfast City

This map is part of an All-Island Atlas project developed by AIRO and the ICLRD. The project is part-financed by the European Union's INTERREG IVA programme managed by the Special EU Programmes Body.

Place of Birth
% Rest of EU27

Less than 1.5%

1.5% to <4%

4% to <7.5%

7.5% to <12.5%

12.5% to <20%

Greater than 20%

Northern Ireland	
Local Authorities	
Water	

Motorways

Trunk/Primary Roads

Secondary Roads

Streets

Ordnance Survey
Ireland/Government of Ireland
Copyright Permit No. MP 005814

Crown Copyright 2014
Ordnance Survey of Northern Ireland
Permit No. 140029

Data Source: Central Statistics Office
(CSO), Northern Ireland Statistics and
Research Agency (NIRSA)

MAP 8.9

Place of Birth - Poland, 2011
Small Areas (SAs)

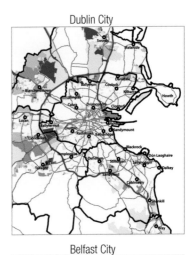

Dublin City

Belfast City

This map is part of an All-Island Atlas project developed by AIRO and the ICLRD. The project is part-financed by the European Union's INTERREG IVA programme managed by the Special EU Programmes Body.

Place of Birth
% Poland

Less than 2%
2% to <5%
5% to <7%
7% to <15%
30% to <30%
Greater than 30%

Northern Ireland
Local Authorities
Water

Motorways
Trunk/Primary Roads
Secondary Roads
Streets

Ordnance Survey
Ireland/Government of Ireland
Copyright Permit No. MP 005814

Crown Copyright 2014
Ordnance Survey of Northern Ireland
Permit No. 140029

Data Source: Central Statistics Office
(CSO), Northern Ireland Statistics and
Research Agency (NIRSA)

Chapter 9

Religion

Dr Andrew McClelland and Justin Gleeson

9.1 Introduction

Any commentary on the religious affiliation of the population of the island of Ireland should be mindful of the potential emotiveness of the subject matter, not least given the historical usage of religion as a proxy indicator for national identity and political allegiance in Northern Ireland. Rather than dwelling on such debates and speculation, the purpose of this commentary is to provide a matter-of-fact overview of the prevalence and geographical distribution of the principal religious groupings identified on the island as derived from the 2011 censuses.

It draws out the key characteristics concerning religion identification and further underlines certain comparisons, within and between, both jurisdictions on the island, with an emphasis on the largest population centres, Dublin and Belfast. Further, a more detailed emphasis is placed on Belfast as a result of the distinct geographical distribution of religion affiliation and community background that exists in the city.

Prior to providing a synopsis of the census data, several issues are worth referencing. Firstly, a number of distinctive questions on religion were posed in the censuses of both jurisdictions on the island. In Ireland, for example, a single question was asked: What is your religion? In Northern Ireland, however, two questions were asked: What religion do you belong to? and, what religion were you brought up in? The differentiated approach in the latter possibly reflects the greater degree of sensitivity associated with religious affiliation in that jurisdiction, and is perhaps also because of the higher rates of those with no religious affiliation. Both questions from the Northern Ireland census are dealt with in this chapter. Secondly, and perhaps more pertinent for a chapter concerning religion, theological questions are not discussed, and it is recognised that the completion of a certain box in the census does not necessarily indicate

personal religious belief or observance of the strictures of that particular denomination. Thirdly, for the purposes of data categorisation at the local scale, only four principal religious groupings are identified:

- Catholic;
- Protestant and Other Christian (including the Presbyterian, Church of Ireland, Methodist and Orthodox denominations);
- Other Religion (Islam, Other);
- No Religion and Not Stated.

9.2 Religious Trends

Trends in religious affiliation in Ireland can be traced for the last 150 years thanks to the recording of detailed census information from 1861 to the current day. A review of historical religious affiliations in both jurisdictions reveals very different patterns, with Ireland being predominantly Catholic with rates in excess of 89% from 1861 to 2011. Northern Ireland – or, at least, the six counties that constitute Northern Ireland - on the other hand, has a more mixed pattern of religious affiliation and has undergone significant transformation within the two dominant religions, Protestant and Other Christian and Catholic, moving from a respective 59:41 split in 1861 to almost parity in 2011.

The dominance of the Catholic Church in Ireland increased from 89% in 1861 to reach its peak in 1961 at a rate of 95%. A gradual decline in the proportional share of the Catholic Church in Ireland began in the period from 1961, falling to 84% (3,861,335) in 2011, its lowest point since records began. The proportional share of the Protestant and Other Christian religions fluctuated little in the period to 1911 when it stood at approximately 10% of the population (Figure 9.1). Post 1911, rates continually declined to a low point of 3.3% in 1991. In the last twenty years there has been a slight recovery, however, with the 2011 proportional share standing at 6.3% (280,330): much of this can be attributed to increasing numbers of Protestant and Other Christian groupings such as Orthodox Christians

(Greek, Russian and Coptic) now residing in the Greater Dublin Area (approx 63% of national total). Ireland also witnessed a large increase in both Apostolic and Pentecostal populations with annualised growth rates of 21% between 2006 and 2011. While the Other Religion population in Ireland has traditionally been very low (less than 1%), recent migration flows from across the World has resulted in the arrival of large numbers of different ethnic identities with associated religions bringing the 2011 rate to 1.9%. For example, large increases are evident in Muslim, Hindu and Buddhist religions in the last 10 years with growth rates primarily located in large urban centres across Ireland. From the 1960s onwards there has also been a noticeable rise in the No Religion/ Not Stated group with rates increasing from 0.2% in 1961 to 7.6% in 2011.

In Northern Ireland, the dominance of the Protestant and Other Christian grouping as a percentage of the population increased from 1861 to the middle of the twentieth century (from around three-fifths to two-thirds), whereas the Catholic population steadily declined to around one-third in the 1920s and 1930s and the early decades of the Northern Ireland state. Although the Catholic population began rising marginally from the mid-twentieth century, with the exception of a period from the early 1960s to the early 1980s, its percentage share of the population in 2011 is virtually the same as in 1861. In contrast, the Protestant and Other Christian grouping has markedly declined from its early twentieth century peak to stand at 41.6% in 2011. In parallel, and arguably partially explaining this almost 25% reduction, has been the rise in the number of those affiliating with no religions or not stating their affiliation, which stood at virtually 0% in 1951 and represented almost 17% in 2011 (10% with no religion; 6.9% not stating). The peak of this latter grouping at 20% in 1981 is explained by the non-cooperation campaign that impacted on the census process, which also accounts for the lowest ever recorded percentage of the Catholic grouping in Northern Ireland that same year (Figure 9.2).

Figure 9.1: Religious Affiliation in Ireland, 1861 to 2011

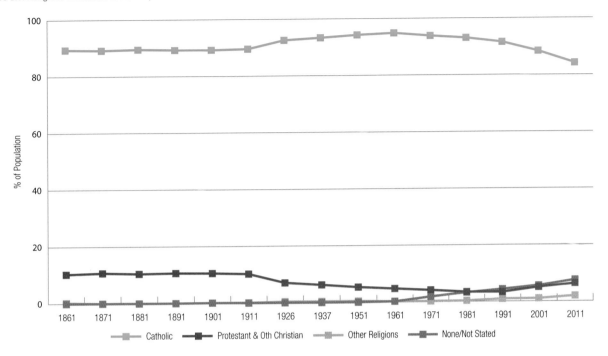

9.3 Religious Patterns in 2011

Taking the population of the island of Ireland into consideration (6.403 million people), the religious breakdown in percentage terms in the 2011 censuses is as follows: *Catholic* (71.9%); *Protestant and Other Christian* (16.3%); *Other Religion* (1.6%); and, *No Religion/Not Stated* (10.2%). It is evident that the large *Catholic* majority on the island is not replicated in both jurisdictions. In Northern Ireland, where a total population of 1.814 million represents 28.3% of the island total, the two largest religious groupings are evenly balanced in their number: *Catholic* (40.8%); and, *Protestant and Other Christian* (41.6%). In marked contrast, 84.2% of the population of Ireland is *Catholic*, whereas the share of the *Protestant and Other Christian* grouping stands at 6.3%. In addition, a further notable difference between the two jurisdictions concerns the higher percentage of the Northern Ireland population identifying with the *No Religion/Not Stated* category. Indeed, the number of those in this grouping is more than double the figure in Ireland, comprising 16.9% versus 7.6% in Ireland.

Nonetheless, even though the *No Religion/Not Stated* grouping represents less than 8% of the population in Ireland, it is the second most populous after *Catholic*. There are also marginally higher proportions of the population in Ireland (1.9%) identifying with *Other Religions* than is the case for Northern Ireland (0.8%). The following section provides further analysis and detailed maps of the distribution of the four main religious categories.

Roman Catholic

The obvious difference between the levels of affiliation to the *Catholic* religion in both jurisdictions, and the immediate fall off in rates just north of the border, is striking (Map 9.1). In Ireland, Catholicism is the dominant religion with 29 of the 34 local authorities containing majorities in excess of 80%. The lowest rates are along the east coast and within the Dublin local authorities. At 75.2%, Dublin City has the lowest percentage, with Dún Laoghaire-Rathdown, Galway City, Fingal and Wicklow all with rates between 76% and 80%. With the exception

of some clusters of low affiliation rates in parts of north Clare and southwest Cork, there is little differentiation in rates across the rest of the country.

The situation in Northern Ireland is far more interesting with very clear and varied spatial patterns across the jurisdiction. At local authority level in Northern Ireland in 2011, the Omagh District Council, Derry City Council and Newry and Mourne District Council areas contain the highest percentage of Catholic populations (greater than 65%), while the east coast council areas of Carrickfergus Borough Council, Ards Borough Council and North Down Borough Council conversely contain the fewest Catholics (less than 12%). Furthermore, certain other local authority areas are more finely balanced between the two largest religious groupings, particularly Craigavon Borough Council – where *Catholic* and *Protestant and Other Christian* each represent 42.1% – Armagh City and District Council – which is 44.8% *Catholic* and 43% *Protestant and Other Christian* – and Antrim Borough Council – where 37.5% of the population is *Catholic*, and

Figure 9.2: Religious Affiliation in Northern Ireland, 1861 to 2011

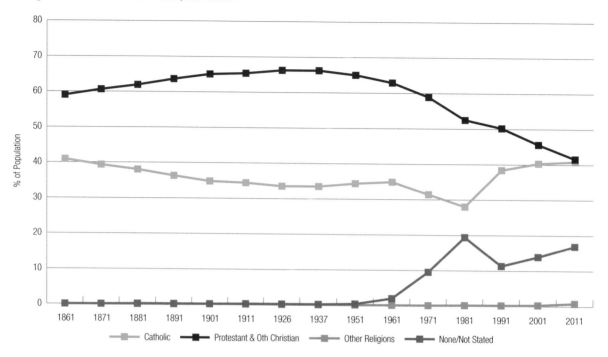

Catholic — Protestant & Oth Christian — Other Religions — None/Not Stated

43.2% *Protestant and Other Christian*. Taken together, 11 of the 26 local authorities in Northern Ireland have a Catholic majority population, whereas *Protestant and Other Christian* forms the majority in a further 15, albeit the extent of the majority is relatively marginal in several cases. A more in-depth analysis of religious and/ or community background majorities across Belfast is provided in the final section of this chapter.

Protestant and Other Christian
The spatial distribution of *the Protestant and Other Christian* religion is largely the reverse of the *Catholic* distribution, given that the population almost exclusively comprises Catholics or members of the *Protestant and Other Christian* religions. Northern Ireland (and to some degree other parts of the north of Ireland) clearly stands out from the rest of the island (Map 9.2).

At local authority level in Northern Ireland, the Carrickfergus Borough Council, Ards Borough Council, Ballymena Borough Council and North Down Borough Council all have rates in excess of 60%, while Newry and Mourne District Council, Derry City Council and

Omagh Distirct Council conversely contain the fewest with affiliations to the *Protestant and Other Christian* denominations (less than 25%).

No such patterns are identifiable in Ireland, however, the highest concentrations of the *Protestant and Other Christian grouping* are evident in Wicklow and the Greater Dublin Region, along the south-east and south coast, and in the border counties of Donegal, Monaghan and Cavan. Nonetheless, no local authority in Ireland contains a *Protestant and Other Christian* population greater than 10%.

Other Religions
The prevalence of the *Other Religion* grouping across the island has increased by 209% in the last 20 years from a figure of 33,328 in 1991 to 103,295 in 2011. At 86% (88,438), the vast majority of the population affiliated to *Other Religions* reside in Ireland, whereas the remaining 14% (14,859) are in Northern Ireland.

The population of those belonging to *Other Religions* on the island of Ireland is primarily concentrated in urban

areas, with the highest percentages in the Dublin local authorities of Fingal (3.7%), South Dublin (3.7%), Dublin City (3.5) and Dún Laoghaire-Rathdown (2.6%). The Dublin local authorities alone account for 59% of the total population living in Ireland. The other main cities in Ireland also have the highest rates and are between 2% and 3%. Interestingly, the two main cities in Northern Ireland - Belfast City and Derry City - have much lower rates, at 1.5% and 0.7% respectively.

In respect of the geographical distribution of the *Other Religion* grouping within the cities of Belfast and Dublin, a very different pattern is evident. In Belfast, the highest rates are in the city centre core. However, in Dublin, the highest concentrations are located in both the city centre and also in other more peripheral locations outside the M50, in areas such as Blanchardstown, Mullhuddart, Lucan and Tallaght (Figure 9.3).

No Religion
Arguably the most fascinating aspect of the 2011 censuses insofar as religion is concerned relates to the significant number of those identified in the *No Religion*

grouping. To illustrate this, the following section looks at this grouping alone and excludes the Not Stated category (Map 9.4).

The highest absolute numbers are unsurprisingly found in the Dublin City Council and Belfast City Council areas (60,081 and 37,239 respectively). Nonetheless, when both these areas are considered in relation to percentages per head of population there are in fact a number of other local authorities with higher rates. At 13.2%, Belfast City Council has the 6th highest rate on the island, and Dublin City Council, with 11.4%, has the 10th highest. With the exception of Dublin City, all of the top ten local authorities in this categorisation per head of population are located within Northern Ireland, predominantly in the greater Belfast area. The highest concentrations are found in the North Down Borough Council (21.3%) and the Carrickfergus Borough Council (18.6%) areas. It is notable that those areas in Northern Ireland that are majority *Protestant and Other Christian* tend to contain the highest percentages in the No Religion category. This is also the case in the Belfast City Council area, where the south and east of the city and the city centre have much higher concentrations of the No Religion grouping. The lowest percentages are identifiably in the west of the city. Interestingly, and at less than a thousand, Strabane Borough Council has the lowest rate of any local authority on the island at 2.4%.

In Ireland, other than Dublin City, the highest *No Religion* percentages are in the urban local authorities of Galway City (10.6%), Dún Laoghaire-Rathdown (10.4%) and Cork City Council (9%), with the lowest concentrations in the border and midlands counties (for example, 2.4% in Cavan and 2.6% in Offaly). There are also some interesting patterns of high rates across parts of northeast Clare, southwest Cork and parts of the upper Shannon region. Within Dublin City there is a tendency for rates to be marginally higher in the south-inner city and along coastal areas in south Dublin.

Whether future censuses will confirm the existence or continuation of a trend towards a more secular population on the island, particularly in urban centres, remains to be seen. This aspect of religious affiliation alone will undoubtedly provide a fascinating focal point for future analyses in both jurisdictions.

9.4 Religious Majorities in Northern Ireland
The Northern Ireland Census also asked an additional question as to the religious denomination or community

that individuals were brought up in. This is slightly different to the direct question on actual current religious affiliation and can be seen more as a measure of community background. The results from this question are therefore also marginally different to the direct religion question. There are two main reasons for this. Firstly, the community background question relates to community brought up in and not current religious affiliation; and secondly, there is an absence of a *Not Stated* (6.75% in religion question) answer in the community background question and as such all answers are either *Catholic, Protestant and Other Christian, Other Religions* or *None*.

Religion or, in this case, community background in Northern Ireland have a unique spatial distribution: large geographic areas have a dominant affiliation to one of the two main religious or community groupings, *Catholics* or *Protestant and Other Christian*. For instance, at the jurisdictional-wide level, an East West differential is evident, with a predominant *Catholic* population in the West and South, particularly along the extent of the border. The *Protestant and Other Christian* grouping is more prevalent in the East and North, with the exception of the Belfast City Council area.

According to the 2011 census, the *Protestant and Other Christian* population is the largest group, at 48.4% (875,717) of the population. Catholics account for 45.1% (817,385); this has increased from 43.% in 2001, which suggests a narrowing of the gap between the two dominant religions. There has also been an increase in those affiliated to other religions or philosophies – now at almost 1% – and a significant increase in those listed as having no religion. This grouping has increased from 2.7% in 2001 to 5.6% in 2011 (an increase of 152%, to 16,592).

There is a very mixed pattern of community background affiliation across the 26 district-council areas in Northern Ireland (Map 9.5). Looking at the percentage of the population with a *Catholic* community background it's clear that some districts, such as Newry and Mourne (79.4%), Derry (74.8%) and Omagh (70.3%), are predominantly *Catholic*, whereas districts such as Carrickfergus (9.6%), Ards (12.7%) and North Down (13.5%), are mainly Protestant.

Much of the rest of Northern Ireland has a far greater mix: the populations of eight of the 26 districts have a *Catholic* affiliation of between 40% and 60%. Areas with

the greatest mix in community background are districts such as Belfast (48.6%), Armagh (48.4%), Craigavon (45.9%), Fermanagh (59.2%) and Antrim (41.2%). Within many of these districts, however, particular areas tend to be dominated by one religious grouping. For example, Map 9.5 indicates the pattern of community background in Belfast. The largest religious grouping in the Belfast City Council area is Catholic, comprising 41.9% of the population, whereas 34.1% are identified as Protestant or Other Christian. The spatial segregation of the residents of the city is more striking than for Northern Ireland as a whole, with the West and East of the city respectively maintaining their *Catholic* and *Protestant and Other Christian* majorities. The North and South of the city are relatively mixed in their composition, although the former consists of a patchwork of concentrated areas of each principal grouping, in contrast to the more gradual distribution evident in the South. Indeed, swathes of the city have proportions in excess of 85% affiliated to a Catholic background. The West has a distinct pattern where predominantly *Catholic* areas, such as the Lower Falls, Turf Lodge and Andersonstown, are separated from Cliftonville and Ardoyne by predominantly *Protestant and Other Christian* areas such as Shankill and Woodvale – the largest 'Peace Wall' also happens to be in this location. The centre and east of the city have a much larger mixing in community background, although the east becomes predominantly Protestant and Other Christian towards the outskirts of the city, in places such as Belmont, Ballyhackamore and Sydenham, and in the suburban local authorities neighbouring the Belfast City Council areas, such as Castlereagh and parts of Newtownabbey.

Areas of the city where the two most predominant religious groupings, *Catholic* and *Protestant and Other Christian*, are close to parity in their number (or within a 10% range), include Finaghy, Windsor and Ravenhill in the South, and Legonial and Fortwilliam in the North. Interestingly, in certain areas in the South of the city where relative parity in number is often assumed, the picture is much more complex. For instance, in the Ballynafeigh district, 44.3% of the population in 2011 is affiliated with the Catholic grouping, whereas 18.4% is *Protestant and Other Christian*. However, 24.9% do not identify with any religion in this area, and a further 12.3% belonging to the *Other Religion* grouping.

MAP 9.1

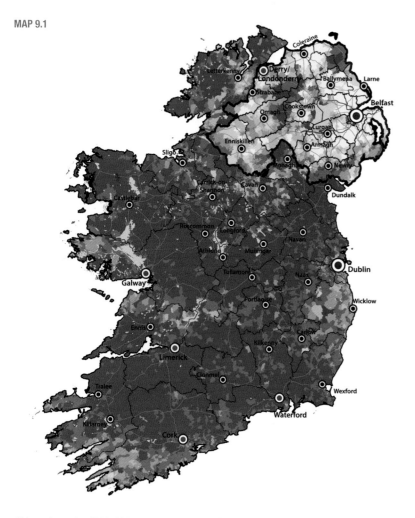

This map is part of an All-Island Atlas project developed by AIRO and the ICLRD. The project is part-financed by the European Union's INTERREG IVA programme managed by the Special EU Programmes Body.

Population with Religion classed as:
Roman Catholic
Small Areas (SAs)

Dublin City

Belfast City

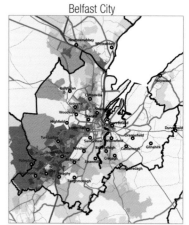

% Roman Catholic

	Less than 21%
	21% to < 46%
	46% to < 65%
	65% to < 78%
	78% to < 88%
	Greater than 88%

	Northern Ireland
	Local Authorities
	Water

	Motorways
	Trunk/Primary Roads
	Secondary Roads
	Streets

Ordnance Survey
Ireland/Government of Ireland
Copyright Permit No. MP 005814

Crown Copyright 2014
Ordnance Survey of Northern Ireland
Permit No. 140029

Data Source: Central Statistics Office
(CSO), Northern Ireland Statistics and
Research Agency (NIRSA)

MAP 9.2

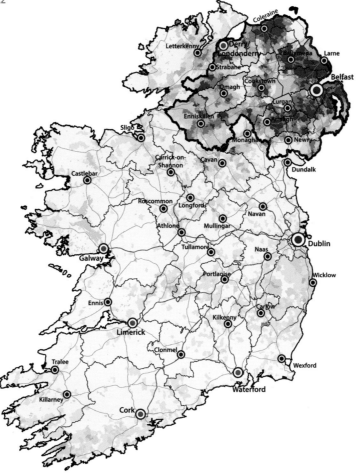

This map is part of an All-Island Atlas project developed by AIRO and the ICLRD. The project is part-financed by the European Union's INTERREG IVA programme managed by the Special EU Programmes Body.

Population with Religion classed as:
Other Christian*
Small Areas (SAs)
*Other Christian is classed as Presbyterian, Church of Ireland, Methodist, Orthodox etc

Dublin City

Belfast City

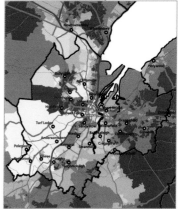

% Other Christian*

	Less than 6%
	6% to < 14%
	14% to < 30%
	30% to < 50%
	50% to < 68%
	Greater than 68%

	Northern Ireland
	Local Authorities
	Water

	Motorways
	Trunk/Primary Roads
	Secondary Roads
	Streets

Ordnance Survey
Ireland/Government of Ireland
Copyright Permit No. MP 005814

Crown Copyright 2014
Ordnance Survey of Northern Ireland
Permit No. 140029

Data Source: Central Statistics Office
(CSO), Northern Ireland Statistics and
Research Agency (NIRSA)

MAP 9.3

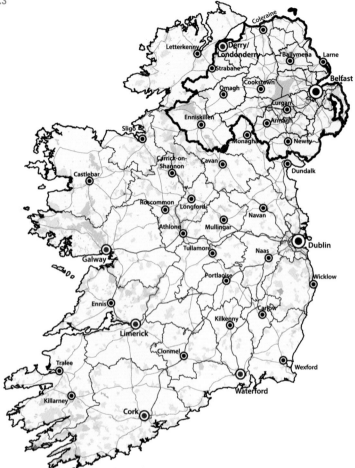

This map is part of an All-Island Atlas project developed by AIRO and the ICLRD. The project is part-financed by the European Union's INTERREG IVA programme managed by the Special EU Programmes Body.

Population with Religion classed as:
Other Religions*
Small Areas (SAs)
Other Religions are classed as Islam, Hindu, Buddist & Other Religions

Dublin City

Belfast City

% Other Religions*

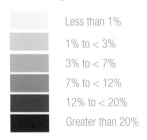

	Less than 1%
	1% to < 3%
	3% to < 7%
	7% to < 12%
	12% to < 20%
	Greater than 20%

	Northern Ireland		Motorways
	Local Authorities		Trunk/Primary Roads
	Water		Secondary Roads
			Streets

Ordnance Survey Ireland/Government of Ireland Copyright Permit No. MP 005814

Crown Copyright 2014 Ordnance Survey of Northern Ireland Permit No. 140029

Data Source: Central Statistics Office (CSO), Northern Ireland Statistics and Research Agency (NIRSA)

MAP 9.4

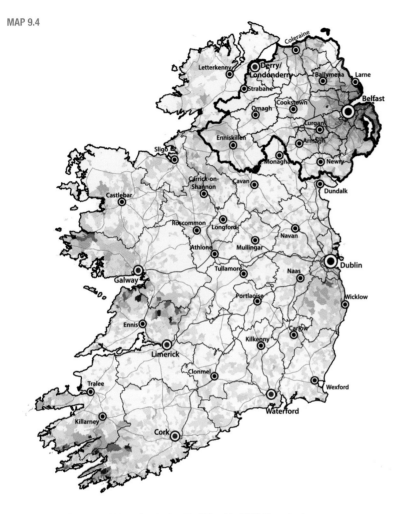

This map is part of an All-Island Atlas project developed by AIRO and the ICLRD. The project is part-financed by the European Union's INTERREG IVA programme managed by the Special EU Programmes Body.

Population with Religion classed as:
No Religion
Small Areas (SAs)

Dublin City

Belfast City

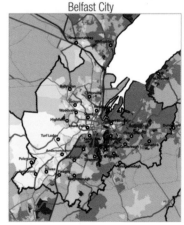

% No Religion

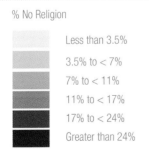

Less than 3.5%

3.5% to < 7%

7% to < 11%

11% to < 17%

17% to < 24%

Greater than 24%

Northern Ireland

Local Authorities

Water

Motorways

Trunk/Primary Roads

Secondary Roads

Streets

Ordnance Survey Ireland/Government of Ireland Copyright Permit No. MP 005814

Crown Copyright 2014 Ordnance Survey of Northern Ireland Permit No. 140029

Data Source: Central Statistics Office (CSO), Northern Ireland Statistics and Research Agency (NIRSA)

MAP 9.5

Religious Majorities in Northern Ireland
Small Areas (SAs)

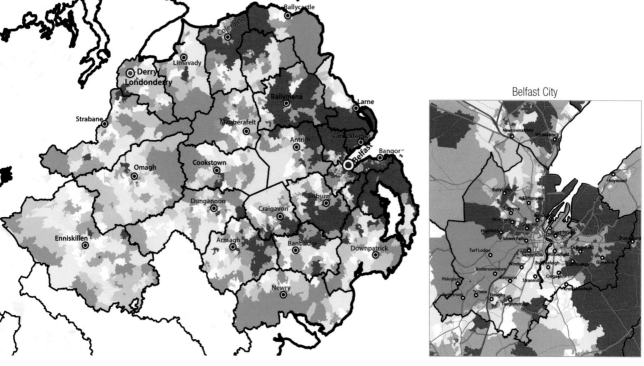

Belfast City

This map is part of an All-Island Atlas project developed by AIRO and the ICLRD. The project is
part-financed by the European Union's INTERREG IVA programme managed by the Special EU Programmes Body.

Religious Majorities in NI

% Catholics	% Protestant
Less than 15%	Greater than 85%
15% to <30%	70% to <85%
30% to <45%	55% to <70%
45% to < 55%	45% to < 55%
55% to <70%	30% to <45%
70% to <85%	15% to <30%
Greater than 85%	Less than 15%

☐ Northern Ireland
☐ Local Authorities
▨ Water

━━━ Motorways
━━━ Trunk/Primary Roads
━━━ Secondary Roads
━━━ Streets

Ordnance Survey
Ireland/Government of Ireland
Copyright Permit No. MP 005814

Crown Copyright 2014
Ordnance Survey of Northern Ireland
Permit No. 140029

Data Source: Central Statistics Office
(CSO), Northern Ireland Statistics and
Research Agency (NIRSA)

Chapter 10

Health and Caring

Dr Ronan Foley

10.1 Introduction

In the absence of individual health records, countries generally measure health and wellbeing in a variety of different and often contradictory ways. Almost all countries record births and deaths (natality/mortality) in a comparable manner. This however is at best an unreliable indicator, especially for meaningful health and social care planning for the living. Data on morbidity is in essence a mapping of illness rather than health but can be used, as health data often is, as a proxy inverse measure. In 2001, the UK Census introduced a new self-reported question on general health, alongside an existing question on limiting long-term illness (LLTI), first collected in 1991. While there were similar concerns about the reliability of such questions, given their subjective nature, they have proved to be good predictors of both mortality and general morbidity (Kyffin, Goldacre and Gill, 2004). In 2011 an EU Regulation (763/2008) promoted the standardisation, as much as possible, of Census questions and a standardised question on general health was introduced in Ireland for the first time to match that asked across all four UK jurisdictions. Questions on informal caring were included in earlier census in both Northern Ireland and Ireland, but were also substantially standardised in 2011.

In asking such a question, the imperative was in part to understand demographic changes and pressures, a particularly pertinent knowledge for the planning and delivery of health and social care. The two jurisdictions do operate different systems for health and social care planning, Northern Ireland following the UK-wide NHS model, while Ireland operates a more hybrid public-private model, within which the Health Service Executive (HSE) currently operate the public service component. Mappable health information is gathered through a number of routes across primary (GP) and secondary (Hospital) sectors. Typically this is gathered by the NHS/HSE, though this only accounts for the public side, and

would typically record where people live and where health care is provided, should they need to access that care. The existence in both jurisdictions of private health care structures means that significant information on health care utilisation remains difficult to access and record. Other specific statistics on mortality (via cancer registries) and morbidity (such as notifiable infectious diseases) are also routinely collected by the relevant government agencies. Finally EU wide and national-level surveys, such as EU-SILC or the Northern Ireland Health and Social Wellbeing Survey, record detailed health status and usage data, but only for small sub-samples of the population.

For the first time in 2011, both populations were asked the same direct question on their self-perceived health status. Respondents were asked to record their health status on a five-point Likert scale running through the tick-box options; Very Good Health; Good Health; Fair Health; Bad Health and Very Bad Health. The rationale for the question was to provide information on the state of the nations' health for a range of spatial scales. As noted, there had been an earlier LLTI question in Northern Ireland in 2001, as well as specific questions in 2001/2 on disability and impairment in both jurisdictions, though these latter statistics were never released in a detailed aggregated or indeed, more generally comparable, form. A parallel semi-standardised question (though with subtle variations North and South) on the levels of informal care, of value to health and social care planners, was also collected in 2011. The role of informal carers has been argued to have an important auxiliary care function, saving state coffers on both sides of the border from more formal health care demand. In knowing something about the detailed geographies of these societal supports, additional knowledge can be gained on wider care demands. The mapped results are discussed in the sections below.

10.2 General Health from the 2011 Census

A joint publication by the CSO and NISRA documented

the headline figures for the general health questions in 2011 (CSO-NISRA, 2014). While a high proportion of the population in both jurisdictions enjoyed good health in 2011, the proportion who considered their health to be good or very good was considerably higher in Ireland, 90.3%, compared to Northern Ireland, 79.5%. At the illness end of the scale, 102,100 (5.6%) persons in Northern Ireland reported their health as bad/very bad compared with 69,700 (1.6%) in Ireland, indicating significantly different levels of perceived poor health. Northern Ireland's results for perceived general health were however, similar to the UK average. In broad geographical terms, the highest rates of self-perceived poor health tended to be found in urban areas, Belfast and Derry in Northern Ireland and Limerick, Cork and Dublin in Ireland. From an age perspective there were expected increases in the levels of bad/very bad health as people aged, with the highest proportions in both jurisdictions being in the over 60 age categories, though with increasing relative levels in Northern Ireland (Figure 10.1).

Given the striking differences in the overall percentages in bad/very bad health between the two jurisdictions (three and a half times higher in Northern Ireland) a number of putative explanations can be offered. First, and most convincingly, the median age in Northern Ireland is 37, compared with 34 in Ireland (the lowest in the EU). The median age across the UK as a whole was 39, so it was clear that the younger population in Ireland, given the data is not age-standardised, would almost by definition, be healthier overall. Equally pertinently, the share of the population over 65 was 15% in Northern Ireland, compared to 12% in Ireland, this typically being the sub-population with the highest levels of morbidity/mortality in general. A second possibility, evident in the rest of the UK and corroborated by LLTI data, was a strong relationship with former industrial areas. Northern Ireland was significantly more industrial than Ireland, especially in relation to the former employment of its now ageing population. Other potential demographic

Figure 10.1: Bad/Very Bad General Health by age, 2011

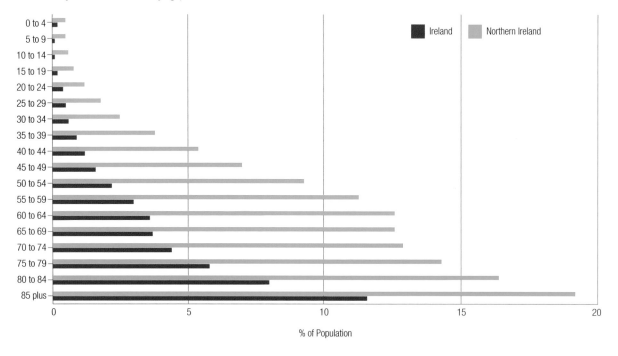

% of Population

and societal differences, such as higher proportions of lone parents and older single person households and a greater level of 'strategic essentialism', a willingness to self-identify as ill or in need of care to access established social supports, may also provide additional explanation.

The following section provides a commentary on the five main categories of perceived general health across Ireland.

General Health Status: Very Good
Map 10.1 displays a small area geography that picks out two broad trends at national level. The first reflects the overall clear differences between Northern Ireland and Ireland, with a much larger number of SAs with high percentage of very good health in Ireland and lower percentages in general in Northern Ireland. In addition, in Ireland there was a general decline in percentages in this category as one moved west to locations with older and more marginalised populations. Commuting zones around the larger cities and towns, typically reflecting younger demographics in such locations, also recorded generally high values for this category, especially in

Ireland. Zooming in to the cities, for both Belfast and Dublin there is evidence of not so much a north-south divide but rather a more diagonal NE-SW divide, reflective in both cities of a generally more affluent population to the right of that line. It should be noted that at city level, the largest concentrations of SAs with percentages of very good health over 60% were found in the more affluent parts of Dublin. By contrast, large parts of West and North Belfast had less than 40% of their populations reporting very good health.

General Health Status: Good
An interesting and slightly unexpected pattern emerges in Map 10.2. Here the percentages were more evenly spread, though still generally higher in Northern Ireland. In Ireland, more rural areas in the West and Midlands had high percentages relative to the South and East. The overall values were lower with a range of values between about 22 and 40%. In the cities of Dublin and Belfast a more complex pattern emerged with slightly higher levels in this category in inner city Dublin and parts of North West Belfast. Overall, the patterns observed on this map were relatable to the previous one for very good health.

It suggested a tendency in Northern Ireland and in the older, more marginalised areas of Ireland, to almost self-consciously choose this more modest appraisal over the 'very good' option. In aggregating the two categories together however, one could begin to identify the parts of the countries where residents in general considered themselves in good overall health.

General Health Status: Fair
This is an interesting category, especially as prior to the 2011 standardisation, the 2001 Census in Northern Ireland, had 'fair' as a middle category in the general health question (the others being Good and Bad). While the percentage of people in this group was smaller - ranging from 6 to 23% - there were still higher percentages across the board in Northern Ireland (Map 10.3). Percentages in Ireland, especially in rural areas, rarely seemed to rise above 10%. In Dublin there were quite low levels generally, though a high percentage in relative terms reported this health status across much of the less affluent parts of West and North Belfast.

General Health Status: Bad

For the category of bad health, the patterns were very similar to those for fair health. General differences between Ireland (generally below 2.5%) and Northern Ireland (generally above 4%) were clear from the mapped results (Map 10.4). The levels within Dublin remain relatively low, while there is an intensification in Belfast on the left hand side of the NE-SW line, picking up the beginnings of a clear association between bad health and other indicators of inequality, such as deprivation and social class.

General Health Status: Very Bad

For this category, there is a general reduction in intensity across both jurisdictions, especially given the much lower percentages who report in this category in Ireland. The geographical patterns at a country and city level are quite similar to those for bad health (Map 10.5), with some clustering in the central rural parts of Northern Ireland. More broadly, and reflective perhaps of the inbuilt associations with age, the lowest rates with very bad health are found in the commuter belts around the cities in Ireland and in Belfast. The highest rates, greater than 6.6%, seem to be concentrated in the centre of Belfast, in the Shankill/Falls districts. One noticeable effect of the Dublin patterns, for both bad and very bad health, is a seeming co-location of small-areas with the highest rates in the vicinity of the major hospitals, which may well be an artefact of the de-facto, rather than usual-residence, reporting of results.

10.3 Kavanagh-Foley Index of Wellbeing (KFIW)

An attempt to present a more cumulative mapping of the health data is presented in Map 10.6. The Kavanagh-Foley Index of Wellbeing (KFIW) was designed to act as a way of capturing the total health of the country, one that could be used to established a general headline health score for individual areas (Kavanagh and Foley, 2014). It was calculated by taking the percentage reporting of each of the five health statuses and weighting them, with a weighting of 1 given to % in 'Very Good' Health and the highest weighting of 5 given to % in 'Very Bad' Health. These were summed together and the resultant score presented what was effectively an inverse index, where the highest scores equated to the poorest level of health. The formula was applied across a range of scales (from regional to small-area) and found a relatively close agreement in means across all of them, with the expected increase in range as you increased the number

of areal units. Like the general statistics, the index is not age-standardised, something that is statistically difficult to do from a single variable combination and at a detailed spatial scale.

The map shows that by combining the individual responses together then the same Northern Ireland – Ireland variation emerges. At SA level the scores range from 112.2 to 283.4 with the mean score of 151.1. A very high proportion of SAs in Northern Ireland are above this mean, with most of Ireland's SAs below it. The areas where an above average score are recorded in Ireland are in more remote rural areas or in city areas that are deprived or close to hospitals. This pattern is less clear in Northern Ireland with a generally above average score across the province apart from affluent Belfast suburbs and wider commuter belts. While the KFIW index does record some good statistical results at broad spatial scales, (an r2 of around 0.6 with ward level deprivation in Northern Ireland), this predictive capacity is much lower at the SA level shown here. The value of the map is its cumulative nature, where the separate patterns observed in Map 10.1 to 10.5 are summarised and visualised in a single map.

10.4 Carers

The 2011 Census showed that there were 214,000 people, accounting for 12% of usual residents, providing unpaid care in Northern Ireland (CSO-NISRA, 2014). This was three times the proportion found in Ireland across all age groups, where 4.1% of usual residents, 187,100 persons, were carers. The joint agencies report suggests that higher proportions of carers in Northern Ireland may again be related to an older age profile and a higher prevalence of self-perceived bad health as discussed previously. Differences in the phrasing of the census questions may also have given rise to different interpretations as to what a carer was. Caring in general was more prevalent in the middle years, typically older adults caring for elderly relatives of friends. For 50-54 year olds for example, 23% were carers in Northern Ireland, compared with 10% in Ireland. In almost all age groups, females were more likely than males to provide unpaid care. Data from 2001 and 2011 showed that the numbers were growing faster in Ireland, but that in both jurisdictions, much of the growth was in the 65 and over age group. One area that has been recorded for the first time, is the presence of young carers, 1.2% of under 15s in Northern Ireland (0.4% in

Ireland) who were known to provide substantial levels of care.

Low Intensity Caring

The first map of carers, Map 10.7, shows the spatial distribution of carers who provide up to 19 hours a week of care. This sub-set contains the largest proportions of all carers, 57% in the North, 54% in the South and generally reflects what might be termed patterns of low intensity caring. As the map shows, the range of values at SA level runs from less than 1.5% to over 10%, with Northern Ireland having the bulk of SAs with more than 5% of its residents engaged in low-level caring. The low levels of care are also evident in rural Ireland and in Dublin, whereas by contrast, the highest levels in Belfast are in the more suburban parts of the city. Evidence from other studies suggest that low intensity caring is more commonly associated with more affluent areas (Foley, 2008).[5]

Medium Intensity Caring

In general, the percentages and proportions of carers reporting in this category, caring for 20-49 hours a week, are lower than the other two, at 17% of all carers in Northern Ireland and 23% in Ireland. There is still a substantial care burden here, but carers seem to self-identify more readily in the other two categories than here. The spatial distribution (Map 10.8) shows that the range of values are lower than for low-intensity caring and also at first glance, seem more evenly spread across the island, though still in general higher in Northern Ireland. In parts of rural Ireland, pockets emerge in the Midlands and along the Atlantic Seaboard while in Dublin, there are clusters in suburban and some inner-city locations. The pattern of an association with the less affluent parts of Belfast also begins to emerge for this category, a sign of that association between income and the need for higher levels of informal care.

High Intensity Caring

The relative proportions of all carers who cared for 50 plus hours a week was 26% in Northern Ireland and 23% in Ireland. These were the carers with the most intense caring responsibilities and it is likely that a large number were effectively operating 24/7. Recorded in the CSO report was the additional finding that the health of carers also diminished with age, in part due to the intensity of this care burden. A very high proportion of this level of care was provided by people in the 35-64

---

5   (Foley, R. 2008): *The Geography of Informal Care in Ireland, 2002-2006.* Irish Geography, *41, 3, 261-278*

age groups, around two-thirds of whom were female. Map 10.9 shows that while the percentages overall were lower, typically around 3%, the levels of high-intensity caring were again higher across the board in Northern Ireland. The city specific patterns again reflected those for medium-intensity caring with a clustering in West and North Belfast. While the levels were lower in Dublin, one might again suggest a possible relationship both here and elsewhere across Ireland of a proximity to care homes.

10.5 Conclusion

While never a substitute for service specific data on access and utilisation of health and social care services, the mapping of general health and care provision across the whole island has real value to policy and service provision. While it will take until the next Irish census to begin to track change, this is already possible in Northern Ireland though a different wording and choice of categories make this still difficult. Data from the UK for example shows identifiable improvement associated with gentrification within cities (ONS, 2013).[6]

 In this atlas the fine-grained patterns of health and caring data do identify some real variations between Northern Ireland and Ireland, in part due to demographic differences in the underlying population, but also reflective of other local factors associated with higher levels of inequality and local cultural variations in relation to the provision of a more specific public health and social care system in Northern Ireland. The constraints linked to comparability of questions make the direct mapping of additionally collected census data on disability and long-term illness difficult and means they do not appear in this atlas. Such data can help to map specific health and social care need and one would like to see a more standardised question on these areas in future censuses, though preliminary evidence suggests they follow similar geographies to those mapped here.

Age-standarisation is inevitably an issue when explaining variations, but is almost impossible to carry out at the micro-scale geographies mapped here. Yet, it is important to recognise the relationship between demand and supply in absolute terms, and the numbers matter here. If one wished to tease out, beyond the effects of age, a more nuanced geography of health, good and bad, for the two jurisdictions; this would make for an interesting piece of modelling work. Yet given that the collection of data on health is still dominated by age, one might argue that a better direction to take would be to work more closely with health care providers at primary and secondary level to produce parallel detailed maps of health service utilisation. A geography of potential health and care demand/need is shown in the maps here, which would really come alive if overlain by data, made available at a reasonably detailed spatial scale, on actual use. But for now, the collection of this data set, in a standardised form across standardised geographies, is a valuable step forward.

6   *Office of National Statistics (2013).* General Health in England and Wales, 2011 and Comparison with 2001. *London. Stationary Office.*

MAP 10.1

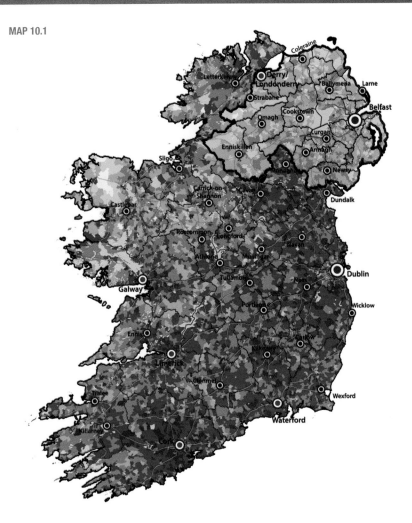

Population with perceived general health as:
Very Good
Small Areas (SAs)

Dublin City

Belfast City

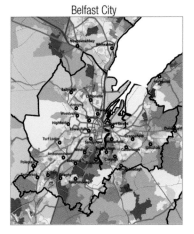

This map is part of an All-Island Atlas project developed by AIRO and the ICLRD. The project is part-financed by the European Union's INTERREG IVA programme managed by the Special EU Programmes Body.

% General Health: Very Good

	Less than 40%
	40% to < 48%
	48% to < 55%
	55% to < 62%
	62% to < 70%
	Greater than 70%

▭	Northern Ireland
▭	Local Authorities
▭	Water

▬	Motorways
▬	Trunk/Primary Roads
▬	Secondary Roads
▬	Streets

Ordnance Survey
Ireland/Government of Ireland
Copyright Permit No. MP 005814

Crown Copyright 2014
Ordnance Survey of Northern Ireland
Permit No. 140029

Data Source: Central Statistics Office
(CSO), Northern Ireland Statistics and
Research Agency (NIRSA)

MAP 10.2

Population with perceived general health as:
Good
Small Areas (SAs)

Dublin City

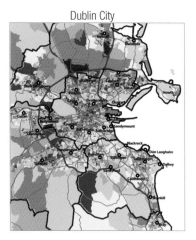

Belfast City

This map is part of an All-Island Atlas project developed by AIRO and the ICLRD. The project is
part-financed by the European Union's INTERREG IVA programme managed by the Special EU Programmes Body.

% General Health: Good

	Less than22%
	22% to < 27%
	27% to < 30%
	30% to < 34%
	34% to < 39%
	Greater than 39%

Northern Ireland
Local Authorities
Water

Motorways
Trunk/Primary Roads
Secondary Roads
Streets

Ordnance Survey
Ireland/Government of Ireland
Copyright Permit No. MP 005814

Crown Copyright 2014
Ordnance Survey of Northern Ireland
Permit No. 140029

Data Source: Central Statistics Office
(CSO), Northern Ireland Statistics and
Research Agency (NIRSA)

MAP 10.3

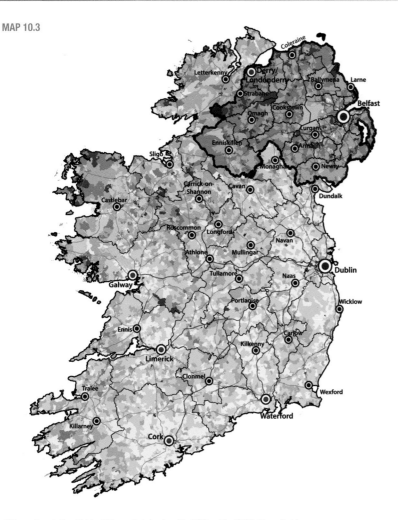

Population with perceived general health as:
Fair
Small Areas (SAs)

Dublin City

Belfast City

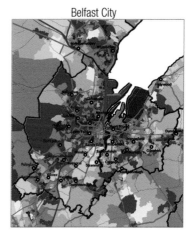

This map is part of an All-Island Atlas project developed by AIRO and the ICLRD. The project is part-financed by the European Union's INTERREG IVA programme managed by the Special EU Programmes Body.

% General Health: Fair

- Less than 6%
- 6% to < 9%
- 9% to < 13%
- 13% to < 17%
- 17% to < 23%
- Greater than 23%

Northern Ireland
Local Authorities
Water

Motorways
Trunk/Primary Roads
Secondary Roads
Streets

Ordnance Survey Ireland/Government of Ireland Copyright Permit No. MP 005814

Crown Copyright 2014 Ordnance Survey of Northern Ireland Permit No. 140029

Data Source: Central Statistics Office (CSO), Northern Ireland Statistics and Research Agency (NIRSA)

MAP 10.4

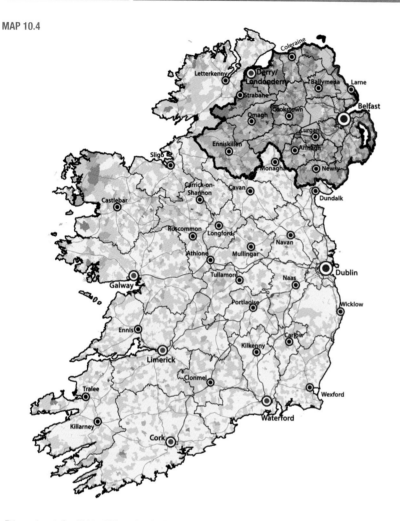

Population with perceived general health as:
Bad
Small Areas (SAs)

Dublin City

Belfast City

This map is part of an All-Island Atlas project developed by AIRO and the ICLRD. The project is part-financed by the European Union's INTERREG IVA programme managed by the Special EU Programmes Body.

% General Health: Bad

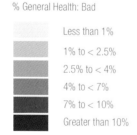

Less than 1%
1% to < 2.5%
2.5% to < 4%
4% to < 7%
7% to < 10%
Greater than 10%

Northern Ireland
Local Authorities
Water

Motorways
Trunk/Primary Roads
Secondary Roads
Streets

Ordnance Survey
Ireland/Government of Ireland
Copyright Permit No. MP 005814

Crown Copyright 2014
Ordnance Survey of Northern Ireland
Permit No. 140029

Data Source: Central Statistics Office
(CSO), Northern Ireland Statistics and
Research Agency (NIRSA)

MAP 10.5

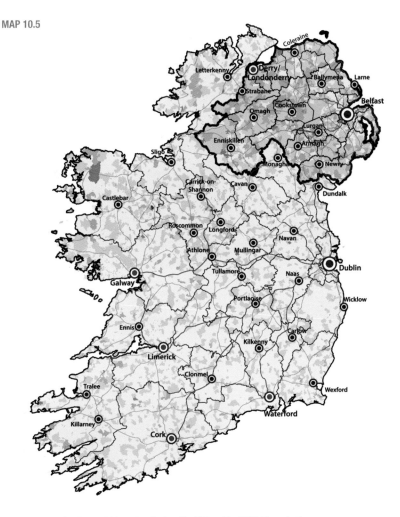

Population with perceived general health as:
Very Bad
Small Areas (SAs)

Dublin City

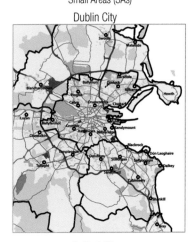

Belfast City

This map is part of an All-Island Atlas project developed by AIRO and the ICLRD. The project is part-financed by the European Union's INTERREG IVA programme managed by the Special EU Programmes Body.

% General Health: Very Bad

- Less than 0.3%
- .03% to < 1%
- 1% to < 2%
- 2% to < 3.7%
- 3.7% to < 6.6%
- Greater than 6.6%

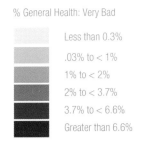

Northern Ireland
Local Authorities
Water

Motorways
Trunk/Primary Roads
Secondary Roads
Streets

Ordnance Survey Ireland/Government of Ireland Copyright Permit No. MP 005814

Crown Copyright 2014 Ordnance Survey of Northern Ireland Permit No. 140029

Data Source: Central Statistics Office (CSO), Northern Ireland Statistics and Research Agency (NIRSA)

MAP 10.6

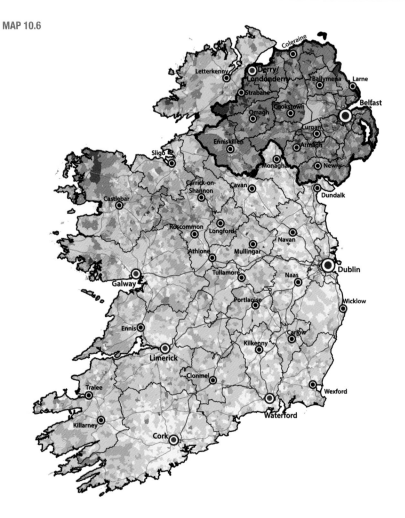

Kavanagh-Foley Index of Welbeing
Small Areas (SAs)

Dublin City

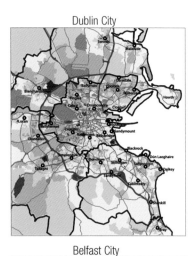

Belfast City

This map is part of an All-Island Atlas project developed by AIRO and the ICLRD. The project is part-financed by the European Union's INTERREG IVA programme managed by the Special EU Programmes Body.

Kavanagh-Foley
Index of Wellbeing

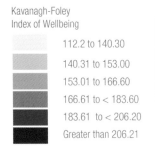

112.2 to 140.30
140.31 to 153.00
153.01 to 166.60
166.61 to < 183.60
183.61 to < 206.20
Greater than 206.21

Northern Ireland
Local Authorities
Water

Motorways
Trunk/Primary Roads
Secondary Roads
Streets

Ordnance Survey
Ireland/Government of Ireland
Copyright Permit No. MP 005814

Crown Copyright 2014
Ordnance Survey of Northern Ireland
Permit No. 140029

Data Source: Central Statistics Office
(CSO), Northern Ireland Statistics and
Research Agency (NIRSA)

MAP 10.7

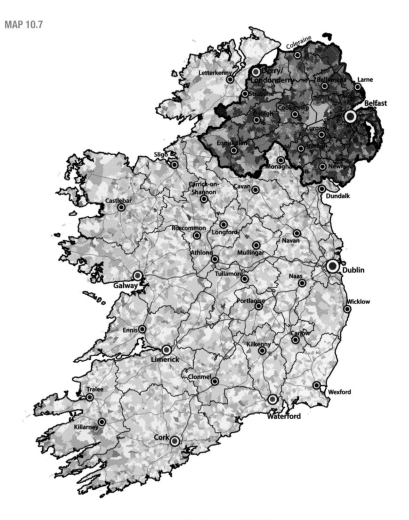

This map is part of an All-Island Atlas project developed by AIRO and the ICLRD. The project is part-financed by the European Union's INTERREG IVA programme managed by the Special EU Programmes Body.

Percentage of Unpaid Workers by hours:
0 to 19 hours
Small Areas (SAs)

Dublin City

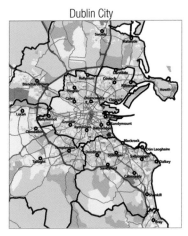

Belfast City

% Unpaid Carers: 0-19 hours

- Less than 1.5%
- 1.5% to < 3%
- 3% to < 5%
- 5% to < 7%
- 7% to < 10%
- Greater than 10%

Northern Ireland
Local Authorities
Water

Motorways
Trunk/Primary Roads
Secondary Roads
Streets

Ordnance Survey
Ireland/Government of Ireland
Copyright Permit No. MP 005814

Crown Copyright 2014
Ordnance Survey of Northern Ireland
Permit No. 140029

Data Source: Central Statistics Office
(CSO), Northern Ireland Statistics and
Research Agency (NIRSA)

MAP 10.8

Percentage of Unpaid Workers by hours:
20 to 49 hours
Small Areas (SAs)

Dublin City

Belfast City

This map is part of an All-Island Atlas project developed by AIRO and the ICLRD. The project is part-financed by the European Union's INTERREG IVA programme managed by the Special EU Programmes Body.

% Unpaid Carers: 20-49 hours

- Less than 0.5%
- 0.5% to < 1.3%
- 1.3% to < 2%
- 2% to < 3%
- 3% to < 6%
- Greater than 6%

Northern Ireland
Local Authorities
Water

Motorways
Trunk/Primary Roads
Secondary Roads
Streets

Ordnance Survey
Ireland/Government of Ireland
Copyright Permit No. MP 005814

Crown Copyright 2014
Ordnance Survey of Northern Ireland
Permit No. 140029

Data Source: Central Statistics Office
(CSO), Northern Ireland Statistics and
Research Agency (NIRSA)

MAP 10.9

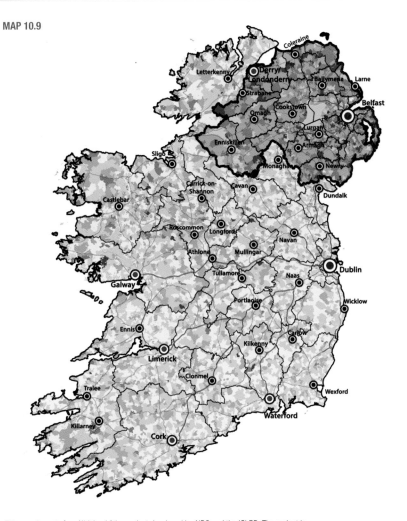

This map is part of an All-Island Atlas project developed by AIRO and the ICLRD. The project is part-financed by the European Union's INTERREG IVA programme managed by the Special EU Programmes Body.

Percentage of Unpaid Workers by hours:
50+ hours
Small Areas (SAs)

Dublin City

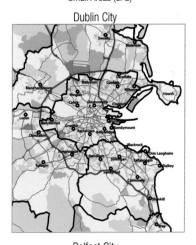

Belfast City

% Unpaid Carers: 50+ hours

- Less than 0.6%
- 0.6% to < 1.5%
- 1.5% to < 3%
- 3% to < 4%
- 4% to < 5%
- Greater than 5%

Northern Ireland
Local Authorities
Water

Motorways
Trunk/Primary Roads
Secondary Roads
Streets

Ordnance Survey Ireland/Government of Ireland Copyright Permit No. MP 005814

Crown Copyright 2014 Ordnance Survey of Northern Ireland Permit No. 140029

Data Source: Central Statistics Office (CSO), Northern Ireland Statistics and Research Agency (NIRSA)

Chapter 11

All-Island HP Deprivation Index

**Trutz Haase, Jonathan Pratchke
and Justin Gleeson**

11.1 Introduction

This study presents an area-based deprivation measure for the island of Ireland based on the 2011 Census. Conceptually, it builds on the study undertaken by the authors in 2011 (Haase, Pratschke and Gleeson, 2012), which used small area (SA) data from Ireland and Northern Ireland to construct a prototype of an all-island deprivation index. The index presented in this chapter embodies further advances, made possible by recent developments in data availability and harmonisation.

The first major development relates to the 2011 Census itself, which was carried out concurrently in almost all European countries, providing data for both Ireland and Northern Ireland. Secondly, small areas (SA) were introduced in both jurisdictions for the publication of aggregate data from the census, providing a better alignment of geographical units. Thirdly, the CSO and NISRA have worked together on an ambitious programme of data harmonisation, leading to a joint publication using socio-economic statistics from the census on an all-island basis (CSO and NISRA, 2014).

The new All-Island HP Deprivation Index builds on these developments and draws on a combined set of equivalent indicators to form a single deprivation index, providing a powerful tool for researchers and policymakers who are interested in understanding and seeking to reduce the social gradient that characterises a multiplicity of different outcomes in the economic, social and political spheres.

11.2 History of Deprivation Measures

11.2.1 Northern Ireland

Deprivation Indices have a long history in both Northern Ireland and Ireland. Indices for Northern Ireland have, without exception, followed the design of those adopted in England. The first deprivation index for Northern Ireland was based on the 1981 Census of Population and was developed to allow the Department of the Environment to identify Urban Priority Areas for targeted interventions under the 1978 Inner Urban Areas Act (DoE, 1983). The resulting index included eight indicators, seven of which were calculated using the 1981 Census. The indicators sought to identify groups known to have a higher risk of poverty, such as lone parents, elderly people living alone and those born outside the Commonwealth.

Following the 1991 Census, a new multidimensional index was constructed, referred to as the Index of Local Conditions (ILC, also known as the Robson Index) (DoE, 1994). The construction of the ILC was guided by hypotheses regarding "domains" of deprivation and differed from its predecessor in making a conceptual shift from the notion of "groups at risk of poverty" to more direct measures, referred to as indicators of "incidence". This new index attracted considerable interest, providing a basis for the designation of eligible areas under successive EU and IFI initiatives to foster peace and reconciliation in Northern Ireland.

The Indices of Multiple Deprivation (IMD, also known as the Noble Index) were the next, in chronological order, to be adopted. The IMD provided new area-based measures for the UK, incorporating eleven separate studies (Noble et al., 2000 to 2007). The IMD differed from previous indices in that they were derived almost entirely from administrative data. The most recent Index of Multiple Deprivation for England (Noble, 2007) is based on seven domains: income, employment, health, education, housing, environment and crime. Each domain comprises a number of indicators, which are combined using the first factor of an Exploratory Factor Analysis (EFA). The domain scores are then combined into a single index score using expert weights. In Northern Ireland, the Index is known as the Multiple Deprivation Measure (MDM)

and was updated in 2001, 2005 and 2010. Index scores are available from the Northern Ireland Statistics and Research Agency.

11.2.2 Ireland

Deprivation indices for Ireland originated with a series of local development programmes that were implemented from the late 1980s onwards, and followed a different trajectory to that described above.

The *Index of Relative Affluence and Deprivation* for Ireland relied on the 1991 Census (Haase, 1996). When compared with the DoE (UK) index based on the 1981 UK Census, this index differs primarily due to its sensitivity to the dimensionality of urban and rural forms of deprivation. In fact, there was a growing perception in Ireland during the 1980s and 1990s that UK indices tended to have an urban bias, given their origin in the designation of Urban Priority Areas and due to the more urban character of UK society.

When developing subsequent indices, Haase and Pratschke introduced an important additional innovation, which was to become the hallmark of the Irish approach to measuring deprivation, namely the estimation of scores which can be compared over time. This approach was first applied in Haase and Pratschke's analysis of data from the 1991, 1996 and 2002 census (Haase and Pratschke, 2005), and subsequently extended to include 2006 data (Haase and Pratschke, 2008; Pratschke and Haase, 2007). This methodology is used in the current Pobal HP Deprivation Index for Small Areas (Haase and Pratschke, 2012). In contrast to the UK, where the census is carried out every 10 years, the Irish Census is repeated every 5 years, providing a stronger incentive to develop deprivation measures that are comparable over time.

The key methodological innovation that allows for comparable scores is the application of a technique known as Confirmatory Factor Analysis (CFA). This methodological technique is quite different to Exploratory Factor Analysis (EFA), which has been widely used in the construction of deprivation indices at international level, but tends to yield different results each time a new set of data is analysed. The method applied by Haase and Pratschke to data from Ireland conserves the positive features of EFA, whilst allowing the underlying dimensions of deprivation to be conceptualised and fixed on theoretical grounds. A range of empirical tests enable the researcher to assess whether the hypothesised model provides an adequate fit to the data, whilst the stability of the measurement structure and scales permit comparisons to be made between index scores relating to different census periods. This combination of a strong conceptual framework, multidimensionality and stable measurement structure provides comparable deprivation scores which are sensitive to the different forms of deprivation observed in different contexts.

11.3 Developing the 2011 All-Island HP Deprivation Index

The 2011 Census is the result of a far-reaching consultation process involving all European countries, and the definitions and procedures used provide the basis for the development of small area deprivation measures at a European level. In the context of cross-border cooperation, the 2011 Census data for Ireland and Northern Ireland provide a unique opportunity to study the spatial distribution of deprivation from a comparative perspective.

This naturally requires a harmonised cross-national dataset, in addition to a statistical model of deprivation that is appropriate in both jurisdictions. Recent UK indices, such as the IMD/MDM, which are based on administrative data, cannot be extended in this way, as equivalent data are not available for Ireland. For this reason, an All-Island HP Deprivation Index (or any supra-national deprivation index in Europe) must be based on the Census of Population.

The development of an all-island index also requires the use of appropriate methodological techniques, which are similar to those used to develop measures which are comparable across time. This was the subject of a previous research project (Haase, Pratschke and Gleeson, 2012), which set out to test the feasibility of extending the statistical techniques used in recent years to estimate deprivation in Ireland to provide comparable scores on an all-island basis. Particular attention

was paid to the following issues: (i) the comparability of indicator variables, (ii) temporal synchronicity, (iii) common dimensionality of deprivation, (iv) fitting a common statistical model, and (v) standardising index scores across multiple jurisdictions.

The main finding of the aforementioned study was that it was feasible to fit a two-group CFA statistical model to data from two jurisdictions, although attention was drawn to the additional work needed to satisfactorily align the raw variables. The current study overcomes all of these remaining obstacles, primarily as a result of excellent work by the CSO, which was able to provide matching census data for all ten indicators used in the construction of the 2011 All-Island HP Deprivation Index.

11.4 Common Indicator Variables Across Multiple Jurisdictions

The 2011 All-Island HP Deprivation Index is unique in being based on a fully-harmonised set of aggregate-level data from two jurisdictions, an exciting precedent in cross-national research and one which would have been considered impossible until just a few years ago. Although the proof-of-concept study referred to above showed that it was technically possible to construct an all-island index – even where there are differences in the way certain variables are constructed – the 2011 index renders these sophisticated techniques largely irrelevant. This lends greater legitimacy to the index, as the variables on which it is based are defined in exactly the same way for Ireland and Northern Ireland. Although

there may be some small differences in how the social class position of families is coded in the two jurisdictions, we believe that these are practically negligible for the variables used here.

Whilst there have never been particular problems with aligning measures of population change or demographic composition, the greatest challenge for harmonisation arose with the measurement of (i) educational attainments, (ii) social class and (iii) unemployment rates. Following considerable work by the CSO in coding and reclassifying variables to match those available for Northern Ireland, we are satisfied that we now have a matched set of indicators for both jurisdictions. The table below provides summary data for these variables, with reference to 23,025 small areas (18,488 in Ireland and 4,537 in Northern Ireland).

11.5 A Conceptual Model of Deprivation

The 2011 All-Island HP Deprivation Index is constructed along the same lines as the *New Measures of Deprivation* (Haase and Pratschke, 2005, 2008) and the *Pobal HP Deprivation Index for Small Areas* (Haase and Pratschke, 2010, 2012), all of which are based on the same set of hypotheses regarding the underlying dimensions of deprivation, and all of which use Confirmatory Factor Analysis (CFA). The index relies on ten variables, each of which expresses a distinct aspect of relative affluence and deprivation. The rationale for the choice of the three component dimensions and their respective indicators is outlined in Haase and Pratschke

Table 11.1: Comparison of Key Variables across Jurisdictions

Variable	Northern Ireland n = 4,537		Republic of Ireland n = 18,488		All-Island n = 23,025	
	Mean	STD	Mean	STD	Mean	STD
Age Dependency	34.4	5.9	32.7	8.2	33.0	7.8
Lone Parents	30.6	19.9	21.5	16.5	23.3	17.6
Low Qualification	42.2	13.3	36.8	15.2	37.8	15.0
High Qualification	22.6	11.5	23.5	14.1	23.3	13.6
Low Class	36.2	14.3	27.4	11.5	29.1	12.6
High Class	29.7	12.9	27.3	13.3	28.0	13.3
Male ILO Unemployment Rate	11.2	7.8	20.1	10.8	18.3	10.9
Female ILO Unemployment Rate	6.1	4.9	12.5	7.1	11.3	7.2
Average Persons per Room	0.45	0.06	0.51	0.18	0.50	0.16

(2005). The dimensions are referred to as Demographic Profile, Social Class Composition and Labour Market Situation, with the following measurement structure:

Demographic Profile is measured by five indicators:
• percentage change in population over the previous five years
• percentage of people aged under 15 or over 64 years of age
• percentage of people with low educational achievements
• percentage of people with a third-level education
• mean number of persons per room

Social Class Composition is also measured by five indicators:
• percentage of people with low educational achievements
• percentage of people with a third-level education
• percentage of households of high social class
• percentage of households of low social class
• mean number of persons per room

Labour Market Situation is measured by four indicators:
• percentage of households with children aged under 15 years and headed by a single parent
• male unemployment rate
• female unemployment rate
• percentage of households of low social class

The final model for the 2011 All-Island HP Deprivation Index fully supports the multidimensional structure of affluence and deprivation that was hypothesised and tested in previous analyses using Irish data. The model fits very well according to the fit indices most commonly used to assess this kind of model (CFI = 0.97; SRMR = 0.04; RMSEA = .08). In order to assess whether the relationships implied by the model are appropriate for both jurisdictions, we also analysed both sets of data separately and estimated a multi-group model with constraints on all free factor loadings to test the invariance of the measurement model (CFI = 0.95; SRMR = 0.08). The only evidence of poor fit in the constrained multi-group model relates to low social class, suggesting that this variable may have a slightly different meaning in each jurisdiction or that it may not be perfectly aligned.

11.6 Interpreting the 2011 All-Island HP Deprivation Index

When faced with the new All-Island HP Deprivation

Measures, practitioners will understandably be interested in comparing these scores with those provided by other indices, and any differences demand explanation. This is not a major issue for Ireland, as the all-island index builds upon previous indices for the Republic and the small area scores for the 2011 All-Island HP Deprivation Index are almost identical to those published for the Pobal HP Deprivation Index for small areas (with a Pearson bivariate correlation of 0.96), which are available online at www.pobal.ie and are widely used by Government Departments and community groups throughout Ireland.

A much greater challenge is posed by the new estimates for Northern Ireland, as the deprivation scores presented here differ significantly from those provided by the Multiple Deprivation Measures. For this reason, it is necessary to discuss some of the differences in how these two indices are conceptualised and constructed. The most important difference is that the MDM, as mentioned earlier, aim to estimate the number of people in each area who are poor. The All-Island HP Deprivation Index, by contrast, conceptualises affluence and deprivation as the extremes of a continuous distribution which reflects not just poverty but also structural weaknesses in socio-economic terms, including disadvantaged rural areas where some of the "would-be poor" have already emigrated.

Secondly, the MDM build exclusively on variables which appear to represent direct and count-like expressions of deprivation. The All-Island HP Deprivation Index, by contrast, is based on the understanding that, when considered at aggregate level, all indicators express the risk or probability that any given individual or family experiences poverty/deprivation. As a result, there is no reason to prefer indicators which are directly correlated with deprivation over indicators which are inversely correlated. For this reason, the All-Island HP Deprivation Index utilises measures of affluence as well as deprivation, including, for example, the proportion of people with a third-level education and members of households which are classified in the higher social classes.

Thirdly, the MDM derive overall scores by summing domain-specific deprivation measures. The All-Island HP Deprivation Index, by contrast, estimates the three underlying ("latent") dimensions of deprivation, which are measured by ten indicator variables, using Confirmatory Factor Analysis. We believe that by adding

domain-specific scores, the MDM involves an arbitrary "double counting" of the same underlying dimensions of deprivation.

Fourthly, the MDM and All-Island HP Deprivation Index result in two very different kinds of distributions. The All-Island HP Deprivation Index is based on a continuous measurement which ranges from extremely affluent to extremely disadvantaged, and the scores are approximately normally distributed (i.e. trace a bell-shaped curve). Measurements are centred on zero (i.e. have a mean of zero) and a standard deviation of ten, which represent true distances from the mean. The MDM, by contrast, aim to capture a one-tailed distribution, without attempting to distinguish between areas characterised by different degrees of affluence. This is clearly less important when using small area maps to identify highly disadvantaged areas (and in this case, both indices perform equally well). It is, however, of great relevance when aggregating to larger areas, as the All-Island HP Deprivation Index generates a population-weighted average across affluent as well as disadvantaged areas, whilst the MDM counts the extent of disadvantage only.

Having constructed a consistent measure of relative affluence and deprivation throughout the island of Ireland, we can now determine where the areas of greatest disadvantage are located and how the two jurisdictions perform in relative terms.

11.7 Mapping and Comparing the 2011 All-Island HP Deprivation Index

After inspecting the scores for the three dimensions that make up the overall index, several observations may be made:

• The ranges of all three dimension scores are roughly similar in both jurisdictions, with the exception of Demographic Growth, which has a higher standard deviation in Ireland, probably reflecting the tail end of population growth and spatially uneven development of residential housing driven by the economic boom
• Mean Demographic Growth is much lower in Northern Ireland than in Ireland (-3.3 compared with 0.8)
• Social Class Composition is, on average, higher in Northern Ireland (1.5 compared to -0.4).
• Most importantly, however, the Labour Market Situation is much more positive in Northern Ireland (8.1 compared with -2.0).

Figure 11.1 Distribution of All-Island HP Deprivation Index Scores

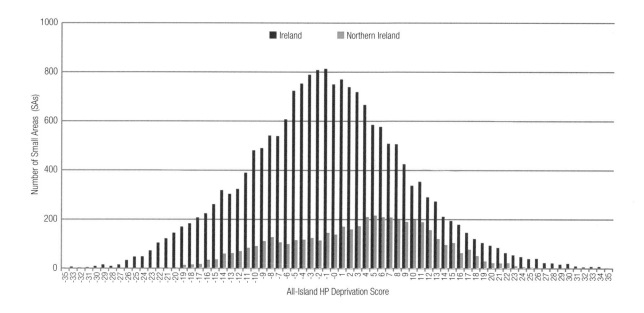

Looking at the combined Index Scores, we can see that:

- Northern Ireland is more affluent, on average, than Ireland (3.0 compared to -0.7)
- Ireland has more extreme values, particularly reflecting areas of severe deprivation (with a minimum of -34.8, compared with -22.1 in Northern Ireland)

Three deeper observations emerge from the current analysis, starting with the obvious result that, by 2011, Northern Ireland had become the more affluent of the two jurisdictions. This is of considerable interest, as the relative positions of Northern Ireland and Ireland are reversed compared with our previous analysis, which was based on the 2001 Census for Northern Ireland and the 2006 Census for Ireland. Table 11.2 below shows the distribution of deprivation scores in the two jurisdictions. The scores for Ireland have a mean of -0.7 and are approximately normally distributed. The mean

for Northern Ireland (3.0) is considerably higher, and the scores follow a bi-modal distribution with shorter tails and fewer highly-deprived areas (Figure 11.1).

Secondly, the driving factor in this striking development has been the ability of Northern Ireland to maintain a comparatively high level of employment despite the unfavourable economic climate since roughly the mid-point of the inter-census period. Ireland, by contrast, has experienced the collapse of the Celtic Tiger and an intensification of economic difficulties and austerity policies. Naturally, these observations relate to 2011, and do not take account of any changes which may have occurred over the past three years in relation to deprivation and its spatial distribution.

The third observation that emerges from this analysis relates to the different ways in which rurality constitutes itself in Ireland and Northern Ireland. In terms of deprivation, the key question is the degree to which peripheral location implies opportunity deprivation, and there appear to be quite marked differences between the two jurisdictions in this regard. In Northern Ireland,

Table 11.2

Variable	Northern Ireland n = 4,537		Republic of Ireland n = 18,488		All-Island n = 23,025	
	Mean	STD	Mean	STD	Mean	STD
Demographic Growth	-3.3	6.0	0.8	10.6	0.0	10.0
Social Class Composition	1.5	9.6	-0.4	10.1	0.0	10.0
Labour Market Deprivation	8.1	9.6	-2.0	9.1	0.0	10.0
2011 All-Island HP Deprivation Score	**3.0**	**8.9**	**-0.7**	**10.1**	**0.0**	**10.0**

people with high educational attainments, prestigious occupations and a high social class position appear to be able to maintain a rural lifestyle whilst nevertheless accessing work-related (and other) opportunities associated with more urban areas. In Ireland, the most important social opportunities (such as accessing more dynamic labour markets) are mainly concentrated in urban areas but, in contrast to Northern Ireland, it appears to be much more difficult to access these from certain areas of the country, presumably due to geographical remoteness, the nature of the transport network and related difficulties.

These factors give rise to a much greater degree of differentiation between urban and rural areas in Ireland with regard to population growth and decline, their ability to retain residents in the central working-age cohorts and their attractiveness to more highly-educated individuals. It would therefore appear that rural areas in Ireland are much more negatively affected by opportunity deprivation than equivalent areas in Northern Ireland. Furthermore,

practically all areas in Ireland have experienced the dramatic and combined impact of economic crisis and cuts in public expenditure, whilst the impact of the crisis in Northern Ireland appears to have been weaker and perhaps somewhat mitigated by the policies that were adopted.

Map 11.1 details the spatial distribution of relative deprivation scores at the SA level across the island. What is clear is the immediate border effect with large parts of Northern Ireland classed as Marginally Above Average and Affluent whereas Ireland is mainly characterised by areas classed as Marginally Below Average and Disadvantaged. The most affluent local authorities/districts on the island are Castlereagh (8.66), Dún Laoghaire-Rathdown (8.5), North Down (8.43), Antrim (6.51) and Lisburn (6.45). With the exception of Dún Laoghaire-Rathdown, Galway City (5.47) is the only local authority in Ireland in the top ten most affluent local authorities/districts. As expected, the most disadvantaged local authorities/districts are all in Ireland

with lowest scores in Limerick City (-8.58), Donegal (-6.06), Wexford (-5.9), Mayo (-5.13) and Tipperary North (-4.6). In Northern Ireland the most disadvantaged local districts are predominantly in the north-east of the jurisdictions with Strabane (-2.75), Limavady (-0.7) and Derry (-0.44) with lowest scores.

An analysis of the two inset maps of Dublin and Belfast illustrates the distribution of deprivation rates across both cities. In Dublin there is a very clear spatial divide with the most affluent areas on the island located in parts of south Dublin and along the coast from Dalkey in the south to Howth in the north. Belfast has a similar spatial divide with the most affluent areas in the central and south-east of the city (Stranmills, Windsor). Interestingly, at the other end of the spectrum Dublin also has areas of extreme disadvantage (Ballymun, Priorswood etc) that are far more disadvantaged than the extremes evident in Belfast (Glencairn, Woodvale, Falls etc).

MAP 11.1

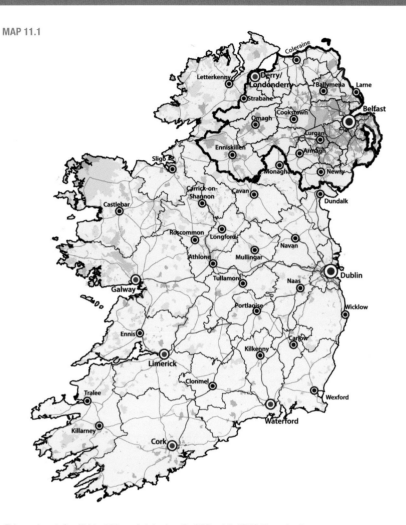

All-Island HP Deprivation Index
Small Areas (SAs)

Dublin City

Belfast City

This map is part of an All-Island Atlas project developed by AIRO and the ICLRD. The project is part-financed by the European Union's INTERREG IVA programme managed by the Special EU Programmes Body.

■ Extremely Disadvantaged		
■ Very Disadvantaged		
■ Disadvantaged		
■ Manginally below Average		
■ Marginally Above Average		
■ Affluent		
■ Very Affluent		
■ Extremely Affluent		

☐ Northern Ireland	▬▬ Motorways
☐ Local Authorities	▬▬ Trunk/Primary Roads
▨ Water	▬▬ Secondary Roads
	▬▬ Streets

Ordnance Survey
Ireland/Government of Ireland
Copyright Permit No. MP 005814

Crown Copyright 2014
Ordnance Survey of Northern Ireland
Permit No. 140029

Data Source: Central Statistics Office
(CSO), Northern Ireland Statistics and
Research Agency (NIRSA)

Conclusion

Justin Gleeson

In 2008 we published the first *Atlas of the Island of Ireland* and noted that although it was a step in the right direction there was still a long way to go before evidence-informed analysis and policy formulation could be undertaken on a routine basis for the whole island. Since then lots of progress has been made and there has been a transformation in the access to, and availability of, spatial data for the island. Much of this progress has been driven by AIRO, the ICLRD and associated research institutes such as NIRSA[1] and the NCG[2] through a long-term research and development programme that aimed at preparing interoperable data across several domains (housing, access to services, health), addressing fundamental technical issues such as incomparable statistical geographies in both jurisdictions, developing new and more sophisticated mapping and analysis tools and educating users (researchers, policy makers, planners) in the use of spatial data in evidence informed planning.

There have been some key developments in the last number of years that have brought us closer to having a complete understanding of the socio-economic characteristics of the island. The first major development relates to the 2011 Census itself, which was carried out concurrently in almost all European countries, providing data for both Ireland and Northern Ireland. Secondly, a collaborative development between the NCG, CSO and OSI resulted in the establishment of the new Small Area (SA) statistical geography for Ireland and as such provided a far better alignment of geographical units north and south of the border. Thirdly, the CSO and NISRA, in cooperation with AIRO, have worked together on an ambitious programme of data harmonisation, leading to a joint publication using socio-economic statistics from the census on an all-island basis. This joint CSO/NISRA publication along with the further

developments undertaken for this atlas has resulted in the creation of the most comprehensive set of all-island socio-economic data ever available. Finally, major advances in on-line GIS software have resulted in on-line mapping viewers now being widely used and allows for the visualisation and interrogation of vast amounts of data (both geographical boundaries and tabular data) in an easy to use manner.

The maps, graphics and accompanying commentary in this atlas provide a fascinating insight into the social and economic characteristics of the island of Ireland in 2011. They reveal that while there are lots of similarities between Ireland and Northern Ireland there are also some key differences with the most striking being in the areas of religion, housing, levels of perceived health and the current economic status of the labour force. The atlas not only reveals clear differences between both jurisdictions on the island but also highlights the very different characteristics of urban and rural dwellers across our island. The development of the All-Island HP Deprivation Index as part of this atlas provides an excellent means of condensing the vast amount of socio-economic data we now have. In summation, the index highlights that, as of 2011, Northern Ireland is more affluent, on average, than Ireland. There are, however, more extreme values in Ireland with both the most affluent and particularly the most disadvantaged areas being south of the border. The index also highlights that in Ireland the degree to which peripheral location implies opportunity deprivation is far more acute than in Northern Ireland.

The development of this atlas has been a tremendously rewarding experience and it is hoped that it will act as a baseline for future policy development and further socio-demographic studies of the island. The overall atlas research project has however had a wider remit

than this hard back publication and we believe that the accompanying on-line digital atlas is as, if not more important than this hard back atlas. The on-line tool provides access to hundreds of variables from a cross section of themes: population, economy, industry, education, transport, housing, nationality, religion, health and deprivation for the 23,025 Small Areas across the island. The mapping interface is an easy to use platform that is accessible to all and requires minimal technical capability to interact with. Throughout the course of this project the team at AIRO have delivered a series of technical workshops, or 'data days', that have focussed on bringing researchers, planners and policy makers from the cross-border area into closer contact with the actual outputs from the atlas. This series of workshops have been very successful and highlighted the importance of developing a further training and mentoring process to ensure a full take-up of the full set of data and tools now available.

Whilst major progress has been made in recent years there are still a number of areas that require attention before there is a full suite of compatible all-island data in place. In the coming year we will see the introduction of postcodes in Ireland and this, dependent on successful integration within central and local government, should herald a new era of administrative data (such as, for example, unemployment benefit claimants etc) being made available on a timely basis at the local level. This, combined with current administrative data already available in Northern Ireland will allow, for the first time, the development of a significant level of local level information that is not linked to the decennial census outputs. Whilst this project has tackled issues relating to socio-demographic data there still remains a major gap in comparative all-island environmental datasets, and attention should be placed on developing all-island datasets and tools focussing on, for example:

---

1   *National Institute for Regional and Spatial Analysis (NIRSA) at Maynooth University*
2   *National Centre for Geocomputation (NCG) at Maynooth University*

flood mapping, water quality, geology and wind energy amongst others. A first step in this process would be the development of an inventory of all-island environmental datasets.

The vast amount of all-island socio-economic datasets now in place are of immeasurable benefit to those interested in cross-border collaboration or issues on an all-island or regional basis, providing a ready means of analysing socio-economic patterns of change or statis and to think through suitable public policy intervention. It is our belief that their creation and availability through this atlas and accompanying analytical tools should be given the highest priority by the government in Ireland and Northern Ireland to enable the highest quality, evidence informed decision making on issues of shared gain.